Principles and Applications of
DIRECT DIGITAL CONTROLS

FI

Ivan Stepnich

Understanding and Implementing DDC Technology
in Refrigeration Applications

Published by

BNP Business News Publishing Company

Troy, Michigan

Library of Congress Cataloging in Publication Data

Stepnich, Ivan C.

    Principles and applications of direct digital controls / by Ivan C. Stepnich.

        p.   cm.

    ISBN 0-912524-47-2 : $29.95 (est.)

    1. Refrigeration and refrigerating machinery—Automatic control. 2. Digital control systems.   I. Title.
TP492.7.S78   1989
621.5 ' 6—dc19   88–30251
                                               CIP

Printed in USA

7 6 5 4 3 2

TABLE OF CONTENTS

DISCLAIMER

HVAC/R INDUSTRY EVOLUTION

By the arrival of the twentieth century, equipment to provide heating, humidification, ventilation and refrigeration was well established. Energy from fossil fuel was readily available and the technology to convert it into steam, hot water, warm air and electricity was well advanced.

By this time, refrigeration could be accomplished in many ways. The absorption principle was discovered by Michael Faraday in the early 1800's. It became a commercial reality in the 1850's and was a mainstay in large refrigeration plants by 1890. Another method, similar to absorption systems (because it did not rely on moving mechanical components), was steam jet or vacuum refrigeration. In addition, the mechanical vapor compression method, the most effective and efficient refrigeration system of all, was also well established and viable at this time. In short, all equipment required for heating, ventilating, air conditioning and refrigeration (hvac/r) was available at the turn of the century.

Despite this progress, environmental technology (especially in the areas of dehumidification and comfort cooling) did not evolve rapidly. Willis H. Carrier, a young engineer just out of Cornell University, took the first positive step toward total indoor environmental control. Carrier's discovery occured on his first project in 1902—the Sackett-Wilhelms Lithographing and Publishing Company of Brooklyn, New York. Excessive environmental humidity was creating serious color printing problems, resulting in unacceptable copy. Willis Carrier made an organized analysis of the problem after a detailed study of humidity and air interaction. This knowledge led to a means of controlled dehumidification which could also be applied to comfort cooling, and hence the air conditioning industry was born.

Later, in 1911, Willis Carrier formalized his research by publishing his "rational psychrometric formula." This formula accurately delineated and correlated the variables of heat energy and moisture in air. This, in turn, resulted in the psychrometric chart which could graphically display all air variables. At last a tool was available for quick reference and rapid, visual solutions to air handling design, application and service problems. Yet despite the new capabilities and adequate equipment technology, comfort cooling or air conditioning for the masses did not become significant until the mid-twentieth century.

Control technology also had an early pioneer. Professor Warren S. Johnson, a Wisconsin school teacher, was one of the first to question the wisdom of a heating system that would periodically overheat or underheat; this was as illogical as operating a vehicle at either full speed or no speed. Professor Johnson experimented with control methods and eventually invented the electric thermostat. The thermostat exercised a method of control by opening and closing dampers. In 1890 Professor Johnson introduced his pneumatic control system, which modulated heating system performance by positioning valves and dampers in proportion to a control signal.

Despite this early success, control technology noticeably lagged behind equipment technology in the first part of the twentieth century. Many refrigeration plants operated with manual control or at best with on/off control methods. Most commercial and residential controls were of the stop/start variety. In addition, the mechanical thermostats had inherent thermal lags which created noticeable temperature overshoots and under-shoots.

This limitation was considerably improved in 1934 by the addition of a heat anticipation device. The heat anticipator added a calculated amount of electric heat to the thermostat's bimetal sensor. As a result, the thermostat responded more quickly in cycling on and off. The ability to cycle off more rapidly eliminated any overheating due to the "flywheel" effect. This improvement served to minimize over-shooting and under-shooting the set point temperature. However, it did not eliminate the problem—it merely made it acceptable. For many years, thermostats provided acceptably inaccurate "saw tooth" performance by staying within the human comfort range.

Electromechanical controllers utilizing wheatstone bridge circuitry were developed in the late 1930's. They provided a control signal proportional to deviation from set point. These were utilized in commercial-industrial applications to modulate valve and damper performance. This ability to match actual performance to signal strength was a great improvement over the "saw tooth" on/off controllers. However, modulating performance did not completely eliminate the constant cycling and hunting characteristics caused by inherent droop and load variables affecting the sensing devices. In addition, the near-linear responses from bridge circuits and pneumatic controllers could not be matched with linear performance from controlled devices because of varying pressure drops through valves and dampers. System design can minimize non-linear performance, but it cannot completely eliminate it.

The 1960's provided the electronic, solid state control systems. Solid state systems permitted greater sensor sensitivity and allowed for system centralization and adjustments at signal centers. More control inputs could be utilized and the multiple sensors could reset or readjust set point to compensate for changing load and weather conditions. In addition, thermistors or wire-wound sensors could be located a considerable distance from the signal center and this could be done with light, unshielded, low voltage wiring.

Solid state systems can only accommodate a limited number of inputs and outputs. In contrast, complete centralized control of refrigeration plants, institutions and high-rise buildings requires a myriad of outputs and inputs in addition to the ability to perform many logic and analysis calculations. The logic capabilities of the solid state signal center are very primitive and limited in comparison with microprocessor-based systems. Clearly, current requirements far exceed the capabilities of solid state signal centers.

Fortunately, arrival of the "space age"—which prompted the development of computers to analyze complex variables and provide scientifically accurate data for space and military programs—was also the turning point for environmental control technology. Computer technology provided logic, memory and software programs capable of handling the complexities of plant and building automation. The control industry suddenly possessed a complete logic unit capable of analyzing the variables in control technology and initiating straight-line set point performance with no perceptible deviation.

At first glance it seemed that the cost of computer hardware and software for control of small refrigeration plants, commercial buildings and even residential application would be prohibitive until well into the twenty-first century. Fortunately, accelerated growth and competition in the computer industry provided the necessary miniaturization and price reduction for broad-based application. This has placed the programmable controller and the microcomputer as the most viable alternatives in environmental control.

In short, the promise of plant and building environmental automation systems in the twenty-first century actually materialized by the end of the twentieth century. At long last, environmental control technology has not only equaled equipment technology, it has surpassed it.

CONTROL
FUNDAMENTALS

Without a brain for guidance and control, the human body cannot function. Heating, ventilating, air conditioning and refrigeration (hvac/r) equipment has the same limitation. A properly designed and applied control system makes the difference between a merely functional system and one that is quite effective and efficient. Consequently, it is important to be familiar with all facets of basic control theory.

■ DEFINITION

A **control device** senses and measures change in environmental conditions or industrial and commercial processes in order to maintain certain conditions within set limits. The conditions to be maintained include temperature, pressure, humidity, air motion, air purity, sound quality, fire safety, building security and energy efficiency.

TERMINOLOGY

Figure 1-1 illustrates the basic elements of a closed loop proportional control system. A sensing control initiates the setpoint command, which is defined as an error signal. Error signal refers to the deviation from a desirable set point. When a deviation occurs, a correction response must be provided by the valve, damper actuator or other control devices. The reaction of the control device is detected and measured, and provides a feedback signal indicating that corrective action is underway.

For better understanding and communication, terminology peculiar to control technology and the hvac/r industry must be defined. The following definitions relate to Figure 1-1 and to all hvac/r control systems:

CONTROLLED VARIABLE: The variable (i.e., temperature, humidity or pressure) which is measured and controlled.

CONTROLLED MEDIUM or **CONTROLLED AGENT:** The substance being controlled (i.e., steam, water, air, fluid or refrigerant).

POWER SUPPLY or **SOURCE OF ENERGY:** The agent which drives the system (systems may be electrical, electronic, mechanical or pneumatic).

CONTROLLER: Unit that detects change in the controlled variable and initiates change in the control device.

Controllers are either **Direct Action (DA)** or **Reverse Action (RA)**. DA refers to an increase in the output signal resulting from an increase in temperature, pressure or humidity (or vice versa). RA refers to an increase in the controlled variable that causes a decrease in the controller output signal (or vice versa).

CONTROLLED DEVICE: The instrument that receives the controller's output signal and regulates the flow of the controlled medium.

SET POINT: The specific value on the controller scale that is to be maintained in the controlled variable.

DEVIATION: The difference between set point and the value of the controlled variable at any instant. The amount of deviation is called the **Error Signal**.

OFFSET: A deviation which is sustained for a substantial length of time.

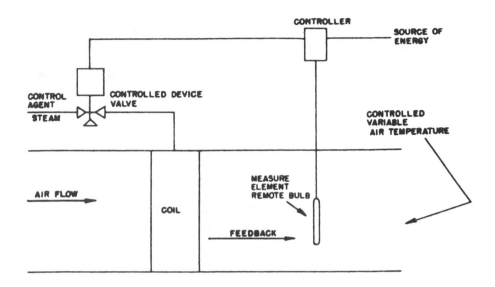

1-1. Configuration of a Basic Control System

DIFFERENTIAL: The difference in the controlled variable which causes the controller to initiate or terminate a signal.

FEEDBACK: Controller action output is returned to the signal origin.

■ CONTROL SYSTEM COMPONENTS

An elementary control system consists of three basic units:

- an energy source
- a controller
- an electrical, electronic, pneumatic, or mechanical operator which positions a control device.

An **energy source** can be electrical, electronic, pneumatic, thermal, vapor pressure or hydraulic. The **controller** senses and measures deviations from the prescribed set point condition and initiates the output which causes the operator to reposition the control device. **Control devices** include valves, dampers, relays, solenoids, pumps, fans, motors, and timing and staging mechanisms.

Solid state programmable controllers and **direct digital control (DDC)** systems accomplish all the foregoing basics. In addition, they exhibit greater sophistication in their ability to make a logical analysis of controller inputs and to provide a mathematically correct response.

SENSORS

The sensor is the fundamental component of the controller. It is also the most important because it reacts directly to changes in the controlled variable. Sensors can react to temperature, pressure or humidity.

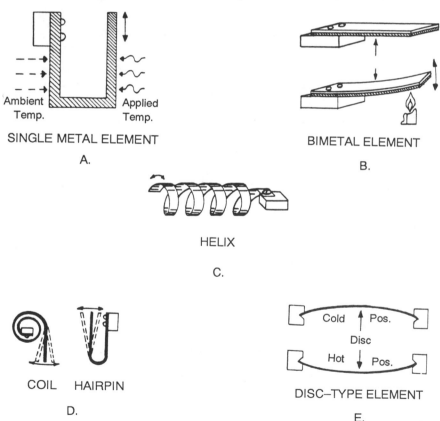

SINGLE METAL ELEMENT

A.

BIMETAL ELEMENT

B.

HELIX

C.

COIL   HAIRPIN

D.

DISC–TYPE ELEMENT

E.

1-2. Variations of the Bimetal Element

Bimetal Sensor

One of the earliest sensors (shown in Figure 1-2) is the **bimetal element**, in which two dissimilar metals are bonded together. Because the metals have different coefficients of expansion, the bimetal strip responds to thermal changes by warping or bending. This bending causes contacts to open and close, or electrical resistance to vary in a potentiometer. Through the years the bimetal sensor has been the predominant component of residential thermostats.

Despite the popularity of the bimetal sensor, solid state and microprocessor technology demanded sensors with more rapid response, greater sensitivity, smaller form and a high coefficient of resistance to change for a given temperature fluctuation (16 to 23 ohms per degree). Even with heat anticipation, the sensitivity of bimetal sensors approaches only one quarter of a degree Fahrenheit (F) under the most controlled laboratory conditions. This means a bimetal thermostat only initiates a controlled response if the temperature changes a quarter of a degree or more.

Electronic Temperature Sensors

Unlike a bimetal sensor, the **thermistor sensor** (shown in Figure 1-3a) can respond to changes as low as 1/260th of a degree centigrade—far beyond the most extreme tolerances of hvac/r equipment. The response of a thermistor is usually a negative coefficient. This means as temperature increases, electrical resistance decreases (and vice versa). Since circuitry design easily compensates for either positive or negative coefficients of change, this is not a disadvantage.

The chief disadvantage of the thermistor is its non-linear tendency, as the curve in Figure 1-3a illustrates. Compensate for this by dial-spacing the degrees on the thermostat to match the thermistor performance. Because this limits the temperature range of thermostats, separate stats are necessary in the low, medium and high temperature ranges.

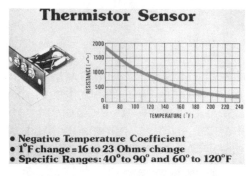

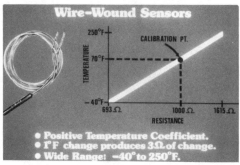

1-3. Electronic Temperature Sensors

The **wire-wound sensor** (Figure 1-3b) also operates on a principle of electrical resistance change per degree of temperature change. And although its rate of 3 ohm resistance change per degree of temperature change is inferior to that of the thermistor, it does have two advantages: a positive coefficient of resistance change, and linearity (as the straight line indicates in Figure 1-4). Because of this, one stat can span a range of -40° F to 250° F.

HUMIDITY SENSORS

The industry has had difficulty in producing reliable and sensitive humidity sensors. For many years, human hair was used extensively as a sensor although its field performance under certain conditions was poor. Other hygroscopic (moisture-absorbing) organic substances such as membrane, animal horn, wood and even biwood (two layers of wood which warped with humidity change) were also used.

The development of nylon-based elements started a new trend towards the use of synthetic sensors. In initial tests these elements displayed excellent laboratory results. However, the contamination buildup on the element in field environments caused problems.

Figure 1-4 shows an improved version of the synthetic humidity sensing elements, called **cellulose acetate butyrate (CAB)**. CAB combines the capabilities of physical change and humidity change, which matches the advantage offered by organic elements. Physical change in CAB can operate electrical contacts or potentiometers, while resistance change is an excellent sensory means for solid state and computer control technology. In addition, the CAB element has an electrical resistance change directly related to humidity change. This positive coefficient of humidity change is approximately 10 ohms per 5% relative humidity.

PRESSURE SENSORS

In addition to temperature and humidity sensing, pressure sensing is a highly essential function in an hvac/r control system. For example, pressures from as low as 2 inches of water in static pressure sensors to more than 400 psi in refrigeration systems must be accurately measured in order to provide optimum system performance and safety. Various pressure sensors are illustrated in Figure 1-5.

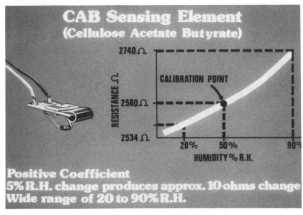

1-4. CAB Sensing Element

Figure 1-5a depicts a metal **diaphragm**, which senses system pressure by exerting an upward force on the diaphragm. Through mechanical linkage this force can also be transmitted to actuate electrical contacts or potentiometers.

Figure 1-5b is a **bellows** pressure sensing device, which provides more significant upward movement than a diaphragm. Both the diaphragm and the bellows produce a force which is calculated by multiplying the gage pressure by the effective area of the bellows or diaphragm.

The **Bourdon tube** is shown in Figure 1-5c. With an increase in pressure, this tube predictably elongates. Through mechanical linkage, it can operate gage needles or electrical contacts. Although it is normally used for pressure gages, the Bourdon tube can be used in operating and limit control.

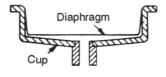

A.   METAL DIAPHRAGM

B.   FABRICATED BELLOWS
OR STACKED DIAPHRAGM

C.   BOURDON TUBE

1-5. Common Pressure Sensors

STRAIN GAGES

The diaphragm, bellows and Bourdon tube are all fundamentally mechanical devices. Even low-range sensors which operate at ultra-low pressures (or even in a vacuum) and read in inches of water (kPa) are mechanical. Sensing elements for low pressure devices can be as simple as an inverted bell immersed in oil, a large slack diaphragm or even a flexible metal bellows.

The **strain gage**, on the other hand, represents the movement from purely mechanical devices towards solid state technology. The strain gage element senses pressure by utilizing a nickel wire or printed circuit resistance mat cemented to a surface under stress. The application of stress (pressure or tension) proportionately changes the electrical resistance of the element. As a result, the electrical signal from a strain gage element reflects the force or pressure as an analog (varying) function.

VAPOR PRESSURE & DIAPHRAGM SENSORS

Although diaphragms and bellows primarily function as direct pressure sensors, they may also be utilized as temperature sensors by vapor charging (bellows) and liquid charging (diaphragms).

Figures 1-6 a & b show a vapor-charged bellows element. To be in control, the bulb must always contain liquid and some vapor. The vapor pressure which develops relates directly to temperature (as any liquid temperature-pressure engineering table can verify).

The one limitation of a vapor pressure sensor is a slight non-linear response (Figure 1-6c). To minimize this limitation, either select a liquid with a nearly linear characteristic in the temperature range required or modify spacing of the temperature indicators on the control dial.

In diaphragm elements (Figure 1-7), the bulb, capillary and diaphragm chamber are all filled with liquid. Any temperature

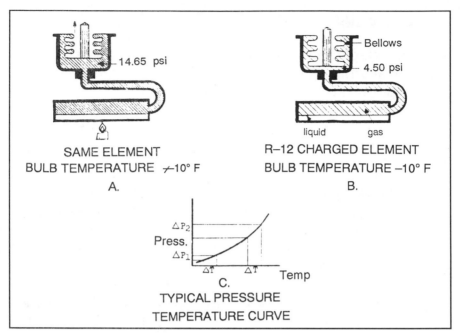

1-6. Vapor Pressure Elements

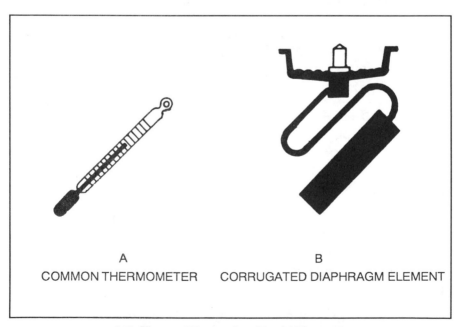

1-7. Thermal Expansion Liquid Elements

change causes a predictable and linear thermal expansion of the liquid, as exhibited by a common thermometer. The upward motion of the diaphragm is not as great as in the bellows, but the force developed is much greater. As with other sensors, linkage can be provided to operate contacts.

Unlike the vapor pressure sensor, liquid expansion sensors are thermally responsive throughout the element. This response area includes the bulb, capillary and the diaphragm chamber. This 'averaging' characteristic is both an advantage and a disadvantage of liquid expansion sensors.

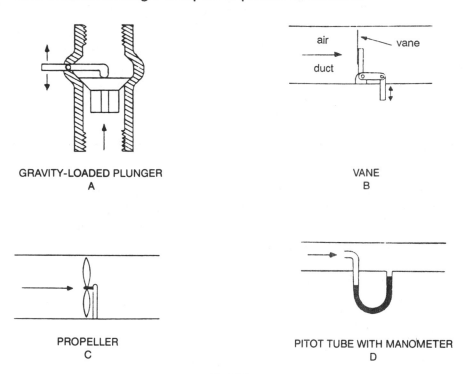

GRAVITY-LOADED PLUNGER
A

VANE
B

PROPELLER
C

PITOT TUBE WITH MANOMETER
D

1-8. Fluid Sensors

'Averaging' is an advantage to limit controls in particular. Limit controls need to sense undesirable local conditions when they occur. For example, a frost condition starting in a small area of an air conditioning coil is damaging if allowed to encompass the entire coil surface. A liquid expansion sensor can

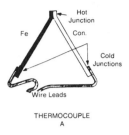

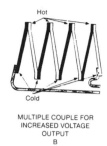

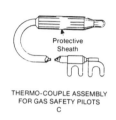

1-9. Thermocouple Sensors

respond thermally to the localized condition and stop the system.

The 'averaging' characteristic is a disadvantage if temperature control is located only at the bulb. In this case the liquid expansion element requires thermal compensation in the capillary and the diaphragm.

FLOW-SENSITIVE SENSORS

Figure 1-8 shows how fluid flow can be sensed. This is an important hvac/r function, as movement of water in hydronic application and movement of air in central fan systems are critical. The vane type is the most popular fluid flow sensor. Deflection of the vane can, through linkage, actuate electrical contacts or potentiometers.

THERMOCOUPLES

Figure 1-9 illustrates a **thermocouple**, a temperature sensing unit which has less application in control than in instrumentation. The thermocouple is based on the physicist

Seeback's observation that dissimilar metals, securely joined, produce electromotive force when heated. The millivolt output of the thermocouple depends directly on the difference in temperature between the hot and cold junctions. Because the thermocouple does not have the mechanical limitations of most conventional temperature sensors, it may see increased usage in the evolution of control technology.

ENTHALPY SENSOR

Air moisture contains a significant amount of heat. The **enthalpy sensor** (Figure 1-10) senses total heat in an air stream and responds to heat content of moisture and heat intensity of air. Using the enthalpy sensor is quite an improvement over the past practice of only using air temperature sensors. It compares total heat in recirculated air with total heat of outside air. This information is very useful when creating a comfortable interior environment or considering energy management systems utilizing outside air. For example, in a central fan application, if outside air is found to have less heat, it can be used for cooling in place of operating costly refrigeration equipment.

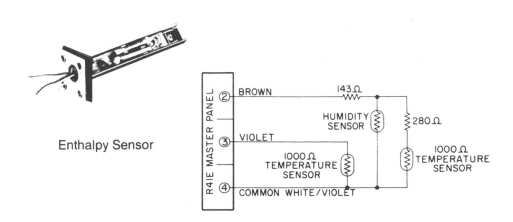

Enthalpy Sensor

Wiring diagram

1-10. Enthalpy Sensor

CONTROL LINKAGE

Transferring an output signal from a sensing controller to the control device requires a linkage system (Figure 1-11). **Linkage** provides contact between the sensor signals and the response device. Linkage is typically electromechanical (Figure 1-11a) or electronic (Figure 1-11b). In DDC systems, the logic system of the microprocessor analyzes the input signal from the sensor and initiates the desired response. Control linkage has four specific functions:

• Power transmission (from sensor to controller device)

• System calibration

• Field adjustment

• Mechanical or electronic amplification of force or motion.

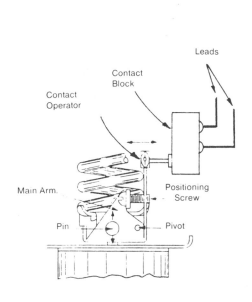

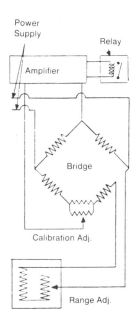

A. ELECTRO-MECHANICAL LINKAGE   B. ELECTRONIC LINKAGE

1-11. Mechanical & Electronic Methods of Control Linkage

CONTROL ACTIONS

Control action from sensor signals provides a pre-determined change in the controlled variable. Control modes are either direct acting or reverse acting.

Direct action (DA) is an increase in the controlled variable (air, water, steam temperature, pressure or any condition being sensed) which causes a direct increase in the control agent (the substance regulated by the control device). For example: if the controlled variable is air in a central fan system and its temperature increase results in an increase of chilled water flow through a valve, this would be direct action.

On the other hand, if a temperature increase in the controlled variable causes a decreased reaction (such as a valve closure or 'throttling down' of the flow of steam or hot water), this constitutes **reverse action (RA)**.

In microprocessor-based DDC systems, the sensing controller conveys its signal as a voltage level. This occurs directly or, in the case of pneumatic controllers, through a transducer. The DDC system software then interprets the signal input voltage via the thermistor as a temperature value.

OUTPUT DEVICES

After proper interpretation of controller signals, the DDC software in microcomputer-based systems must be programmed to provide precise performance from output devices. It is the output devices that cause hvac/r equipment such as fans, pumps, valves, dampers and chillers to perform according to design requirements. Output devices can be pneumatic, electromechanical or electronic.

Pneumatic Output Devices

Output devices are responsible for performance demanded by the controller. These devices include relays, solenoids, motor starters, sequencers, timers or valve and damper actuators.

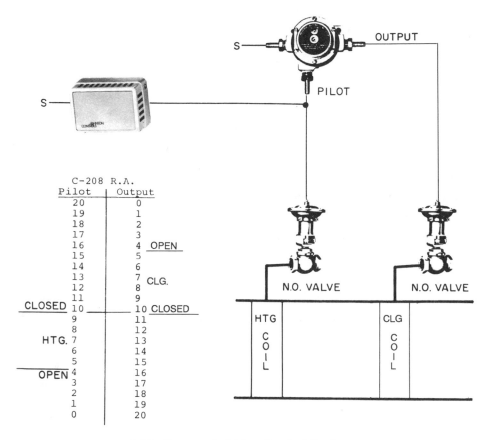

```
     C-208 R.A.
    Pilot  |  Output
       20  |    0
       19  |    1
       18  |    2
       17  |    3
       16  |    4  OPEN
       15  |    5
       14  |    6
       13  |    7  CLG.
       12  |    8
       11  |    9
CLOSED 10  |   10  CLOSED
        9  |   11
        8  |   12
HTG.    7  |   13
        6  |   14
        5  |   15
OPEN    4  |   16
        3  |   17
        2  |   18
        1  |   19
        0  |   20
```

1-12. Pneumatic Valve Control System
Courtesy of Johnson Controls, Inc.

Figure 1-12 illustrates pneumatic control devices involving valve application. A normally open (NO) valve is used in the heating mode. In the absence of any control pressure or energy the valve is open. A direct action (DA) controller operates the valve. As the controller senses increased temperature, it increases pneumatic pressure. This reaction 'throttles down' the output of steam or hot water. With lower temperature, pneumatic pressure is decreased and the valve starts to open. This increases both the flow of the steam or hot water and the temperature of the controlled variable.

In the cooling mode, the valve is also an NO valve but the controller is reverse acting (RA). As temperature increases,

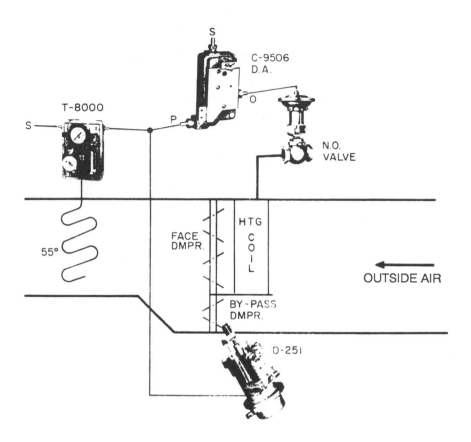

1-13. Pneumatic Valve & Damper Control System
Courtesy of Johnson Controls, Inc

pneumatic pressure decreases to allow more chilled water through the coil to cool the controlled variable. With a temperature decrease, pressure increases to drive the NO valve towards the closed position.

Pneumatic controllers and actuators can function with Direct Digital Control systems if conversion devices are used. A pneumatic-to-electronic transducer converts a pneumatic signal to an electrical or electronic signal that the microcomputer can interpret. And with the help of an electric-to-pneumatic transducer, the microcomputer's output response is converted to pneumatic pressure for the output actuator.

Modern DDC technology provides transducers and interface modules for practically any transition requirement.

Figure 1-13 shows a typical air handling system. A face and bypass damper actuator is used in conjunction with a heating valve actuator. If heating is required, the damper actuator causes the face damper to open and the bypass damper to close. If cooling is required, the face damper closes and the bypass opens.

Electromechanical & Electronic Output Devices

Electromechanical and electronic actuators are similar to pneumatic devices in their ability to operate valves and dampers in proportion to controller demand.

Figure 1-14 illustrates electrical motor actuators which have limited rotation (180° or less). The motors can be on/off two-position or proportional. The output shaft of the motor in Figure 1-14a can be fitted with a damper linkage assembly and crank arm for damper operation.

1-14. Electric Motor Actuators

Figure 1-14b is fitted with a linkage for complete valve operation.

Figure 1-14c can be fitted with a switch kit which allows staging operation or multi-stage refrigeration machines, fans or boilers. The motor actuators can also have additional potentiometers (either built-in or external) to operate other actuators in a master-slave configuration.

■ SUMMARY

Hvac/r equipment requires control devices to sense and measure change in the controlled medium. Various types of sensors are used, depending on the system and the medium being controlled. Sensors can measure variables such as temperature, humidity, pressure, fluid flow and air heat. Linkage systems are used to transfer sensor signals to the control device. Output devices interpret these signals and initiate a proper corrective response in a control system.

CONTROL
SYSTEMS

Control systems in the twentieth century evolved from relatively primitive performance to abilities of undreamed sophistication. Space age requirements demand control systems with microsecond accuracy—a millisecond error in space can be devastating. Similarly, direct digital control systems give the hvac/r industry precise centralized control capabilities which meet the strictest requirements of any application.

However, not all applications require a high degree of sophistication. Consequently control system designers need to evaluate all systems and determine which is the most appropriate and cost effective for each application.

■ CONTROL ACTIONS

Control systems are based on six basic **control actions**:

- Two position (on/off)
- Floating control
- Proportional control
- Proportional plus Integral (PI)
- Proportional plus Derivative (PD)
- Proportional plus Integral and Derivative (PID)

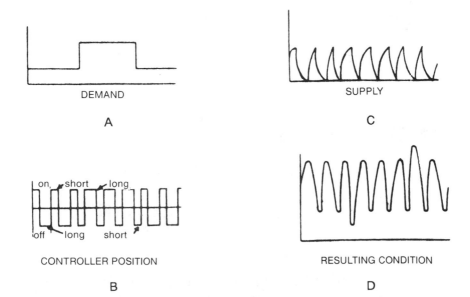

DEMAND

A

SUPPLY

C

CONTROLLER POSITION

B

RESULTING CONDITION

D

2-1. On/Off (Two Position) System Cycling Characteristics

ON/OFF SYSTEMS

The **on/off system** (diagrammed in Figure 2-1) is either fully energized or fully de-energized. The system provides a set point adjustment for the controlled variable. The control differential is the deviation from set point which initiates action from the control device. This deviation can be fixed in some controls and is only dependent upon the mechanical or electrical characteristics of the control linkage. Limit controls, which function only when undesirable conditions develop, usually have fixed differentials. In contrast, operating controls—which perform the required equipment cycling—usually have adjustable differentials.

Although these adjustments allow the system to perform more precisely, on/off control systems do have limitations. An on/off system maintains conditions only within a range dictated by the differential. As a result, it constantly cycles within this range (indicated by Figure 2-1b). This limitation of precision does not suit the strict tolerances of modern equipment. Consequently on/off control, which was predominant in the early twentieth century, is being replaced by more advanced proportional controllers. However, on/off systems can work satisfactorily where the ratio of load requirement to system capacity remains constant.

FLOATING CONTROL SYSTEMS

The **floating control** (Figure 2-2) contains elements of both the on/off and proportional (modulating) control methods.

Modulating performance is achieved by a reversing motor which drives a valve or damper to full open or full closed, or allows it to float at intermediate settings. When the thermally-actuated moveable contact (Figure 2-2e) touches a stationary contact, the motor is driven to a new position (which may be intermediate). It floats in this null position until the moveable contact again touches one of the two stationary contacts.

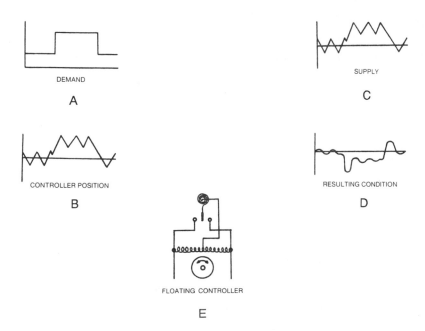

2-2. Floating Control System Cycling Characteristics

As Figures 2-2b and d illustrate, when the set point is at an intermediate position of the moveable contact, the system cycles between the fully open and fully closed positions of the control device. This method's accuracy depends upon the differential setting. A narrower differential provides greater accuracy. However, it must not be so narrow that rapid cycling or hunting results.

This elementary system of proportional control works better in theory than in practice. True proportional control systems are much more effective and practical. In some cases the cost advantage of a floating control system may be a factor, but most installations require performance standards attainable only with a true proportional control system.

PROPORTIONAL CONTROL SYSTEMS

The **proportional control system** is referred to by various names: gradual or throttling action, proportional position action, conversion response or—more commonly—modulating control. It derives its name from the ability to produce an output signal or response directly related to the controlled variable. In control terminology, the final controlled element position is a linear function of the controlled variable. This is mathematically depicted by:

$(Y-Yo) = -K(V-Vo)$

where:

Y = The final controlled element

Yo = The final controlled element position

K = Constant

V = The controlled variable value

Vo = The controlled variable initial value

To maintain the controlled variable at desired value, the set point of the controller dictates the position of the control device. However, when loads change, a sustained deviation from set point must occur in order to reposition the control

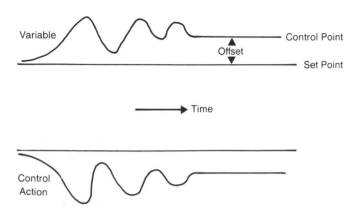

Variable

Control Point

Offset

Set Point

Time

Control
Action

2-3. Proportional Control Performance Characteristics

device. This deviation is termed offset, drift or droop. Thus the actual control point may not be simply the set point, but the set point plus or minus the offset (shown in Figure 2-3).

The amount of change in the controlled variable which causes the control device to move from one extreme to another is called **proportional band** or **throttling range.** Proportional controllers usually have this type of adjustment on the dial for matching system performance with load requirements.

In a well-designed and properly adjusted system the control device delivers an output in proportion to the controller's error signal. A feedback loop from the control device to the controller either pneumatically, electrically or electronically cancels the error signal as the controlled variable attempts to achieve design conditions. The control device constantly repositions itself to provide the proper amount of controlled agent (i.e., chilled or heated air, steam, hot or chilled water) according to the controller's demand.

Figure 2-4 illustrates proportional control in an electromechanical system. Controller T1 has a temperature-sensitive element which drives a potentiometer's moveable

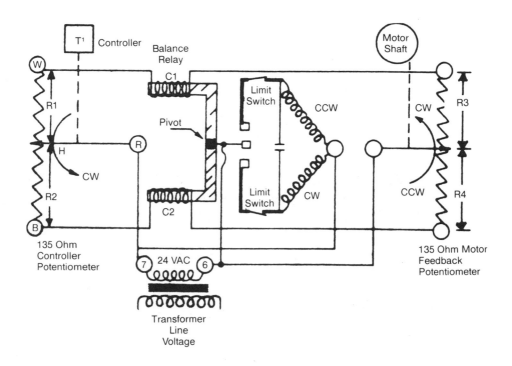

2-4. Proportional Control: Balanced Bridge Circuit

finger from one extreme position to the other. The moveable finger then contacts a wire-wound resistance element. As the finger moves, resistance and current in one leg of the bridge circuit vary, as do resistance and current in the other leg. This action energizes the appropriate clockwise (CW) or counter-clockwise (CCW) motor windings. In this manner the control device repositions to meet load requirements.

The other mechanical and electrical components in Figure 2-4 occupy a proportional motor which moves an output shaft forward and reverse 180°. This shaft drives a valve or damper from fully open to fully closed. The shaft also controls differential end switches for staged operation from a no-stage mode to all-stage operation. The transformer shown is usually an integral part of the proportional motor but is sometimes a separate component.

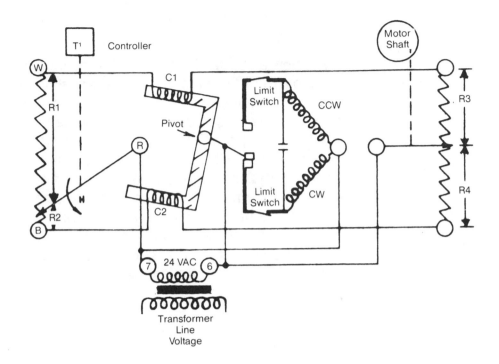

2-5. Proportional Control: Unbalanced Bridge Circuit

The temperature controller in Figure 2-4 is at set point, the control device is at mid-point and the controlled variable is at desired value. No change occurs because the electromechanical version of a wheatstone bridge circuit is in balance.

The moveable fingers on the controller and the motor's feedback potentiometer divide the bridge circuit into an upper leg and a lower leg. The upper leg contains resistor R1 in the controller and R3 in the feedback potentiometer. The lower leg contains resistors R2 and R4. The resistance values in the upper leg and lower leg are equal. Consequently (and according to Ohm's law) the currents running through the balance relay coils C1 and C2 are also equal. Because these coils pull with equal force on the pivoted horseshoe armature, the moveable contact is at mid-point. In this mode the CW and CCW windings of the proportional motor (which operates the control device) cannot be energized. As a result the controlled variable is at set point and the control system is satisfied and in balance.

Figure 2-5 shows how a sudden increase in the controlled variable causes the controller's finger to pivot to its lowest position. This movement causes the upper leg of the bridge circuit to increase resistance. Since electrical current takes the path of least resistance (Ohm's law) coil C2 becomes stronger than C1. This tilts the armature so that the moveable contact strikes the fixed contact and energizes the proportional motor's CW windings. This clockwise motion causes the feedback potentiometer's moveable finger to move upward to rebalance the resistance in the upper and lower legs of the bridge circuit. The proportional motor reacts by positioning the control device in proportion to a signal from the controller (Figure 2-6).

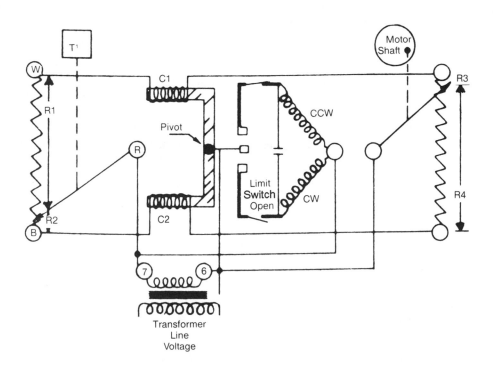

2-6. Proportional Control: Rebalanced Bridge Circuit

Figures 2-5 and 2-6 illustrate an extreme thermal change in the controlled variable. Extreme change requires an extreme response by the control device. As the extreme thermal load conditions subside, the controller repositions itself and causes the proportional motor to reposition itself by rebalancing the bridge circuit. In this manner, the control device always positions itself in accordance with environmental load conditions. However, in most well-designed systems the changes are not usually this drastic, as exhibited in the example of an electromechanical method of proportional control presented in the following section.

Figure 2-7 is an example of a solid state proportional control system. The controller consists of a thermistor and set point adjuster in one leg of a wheatstone bridge circuit. Deviation from set point unbalances the bridge and creates a circuit from

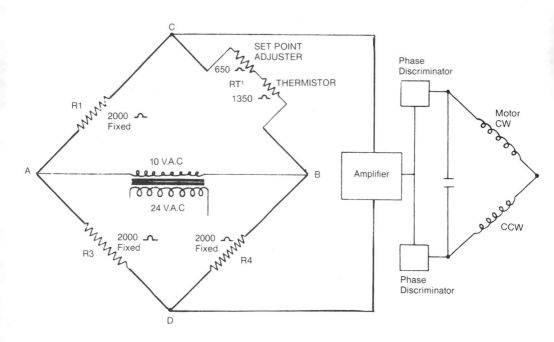

2-7. Wheatstone Bridge Principle

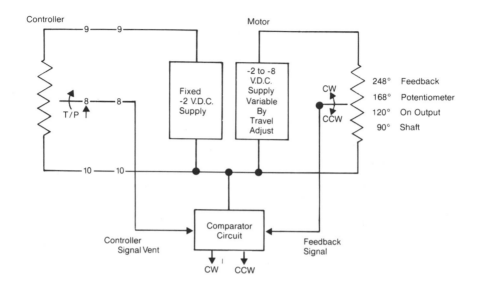

2-8. Solid State Proportional Circuit

point C to point D. This signal is amplified and the phase discriminators determine whether it is a call for heating or cooling (depending on the phase characteristic of the signal). This determines which winding of a proportional motor will be energized. The final position of the motor is again dictated by a feedback potentiometer on the output shaft of the motor.

A more modern circuit is shown in Figure 2-8, where (similar to Figure 2-7) a comparison circuit determines the direction and degree of the controller's demand signal and positions the control device accordingly.

Pneumatic Proportional Control Principles

Electromechanical and electronic control systems accomplish proportional control with relative ease. The controller signal and the control device response (through potentiometers, thermistors and various solid state sensors) relate directly to electrical resistance change. By utilizing bridge cir-cuitry, the resistance changes can actuate control devices and accurately position them to meet the precise requirements of the control system. In contrast, pneumatic systems—which are non-electrical and rely on air pressure of 15 to 20 psig as an energy source—must accomplish proportional control mechanically.

Figure 2-9 illustrates a basic pneumatic system. An air compressor supplies system pressure which is regulated and reduced by a reducing valve. The controllers require the proper pressure to run control devices. This modulated controller pressure, in conjunction with a back pressure feedback arran-gement, proportionately positions the pneumatic actuator. The feedback principle designed in the controller mechanism is demonstrated in diagrams 2-10—2-12.

Figure 2-10 illustrates a pneumatic temperature controller providing proportional action. The bimetal temperature sensor leaves the control port completely uncovered so the system is in the off position. In this mode, the main valve prevents air from entering the lower diaphragm chamber and traveling through the control air line. As a result, air pressure is not sup-

A Simplified Basic Control System. Air Compressor (Power Supply) Supplies Pressure to Room Thermostat (Controller) Which Sends Output Pressure Signal to Damper Actuator to Position Damper (Controlled Device).

2-9. Basic Pneumatic Control System

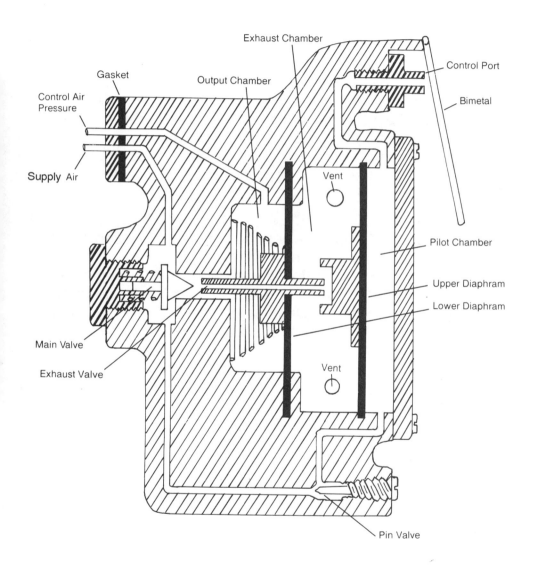

2-10. Pneumatic Temperature Sensor

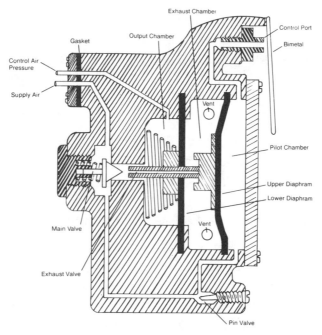

2-11. Pneumatic Temperature Sensor, Intermediate Position

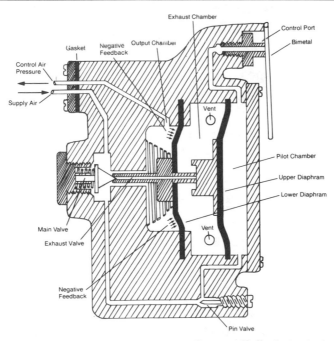

2-12. Pneumatic Temperature Sensor, Fully Actuated

plied to the control device. However supply air does pass through the main valve, the adjustable pin valve, the pilot chamber and then (in the off mode) exhausts or "bleeds" to the atmosphere. No air flows to the control device because air in the output chamber is allowed to only pass through the exhaust valve into the exhaust chamber where it is vented.

Figure 2-11 illustrates a situation in which the bimetal sensor reacts to change in the controlled variable. The bimetal sensor starts to restrict the control port, which in turn creates a pressure buildup in the pilot chamber. This deflects the upper diaphragm to the left and starts to close the exhaust valve.

In Figure 2-12 the bimetal sensor has completely closed the control port. Pressure in the pilot diaphragm is full system pressure and the upper diaphragm is deflected to the left, completely closing the exhaust valve. The main valve is also pushed off its seat. This allows full supply air pressure to enter the output chamber and control air pressure line and then travel to the control device. As diagram 2-12 shows, when a bimetal sensor senses an extreme condition it places the control device in a fully open position.

In a pneumatic system, the feedback which proportionately positions the control device is called **negative feedback.** It consists of the back pressure on the lower diaphragm (represented by the small arrows). This negative feedback pressure builds up until it matches the pressure in the pilot chamber.

If the bimetal sensor decreases pressure in the pilot chamber, the upper diaphragm deflects to the right. This closes the main valve and opens the exhaust valve, which bleeds air from the exhaust chamber. This allows the control device to assume a new position. The air pressure in the output chamber decreases only until negative feedback matches the pressure in the pilot chamber. Therefore as pilot pressure increases or decreases in relation to the thermal condition of the controlled variable, negative feedback proportionately changes to maintain setpoint requirements.

Pneumatic On/Off Controller

Two position or positive on/off action in pneumatic controllers requires a more complex mechanism than does proportional action. Two position action requires an additional component: a positive feedback chamber (shown in Fig. 2-13, component #5). This chamber is required to drive the main valve to its fully open position whenever the sensor triggers an error signal.

Figure 2-13a illustrates two-position action. The controller initiates a signal which begins to pressurize the pilot chamber. Deflection of the pilot chamber diaphragm causes the main valve to open, which allows supply pressure to output to the control device. Also, supply pressure enters the drilled passage in the area of the output pressure outlet. This passage leads to the positive feedback chamber; the added pressure then drives the main valve to the fully open position. This places the control device in its maximum 'on' position.

When the error signal is cancelled, the main valve closes and the exhaust valve opens. This allows both output supply pressure and pressure in the positive feedback chamber to fall to zero. As a result the control device returns immediately to the closed position.

Open and Closed Loops

In the previous examples, feedback provided a means for true proportional action. The presence of a feedback loop assures the controller that the control device's action equals the magnitude of the error signal. The feedback loop can be mechanical, electrical, electronic or a combination of the three. In system application, feedback from the controlled variable to the controller determines whether the loop is defined as open or closed.

The **open loop system** (Figure 2-14a) contains an outdoor temperature sensor which controls heat input to the room. Heat input (Hi) is regulated exclusively by outdoor temperature (To); room temperature (Rt) does not affect the amount of heat

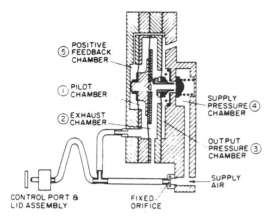

A: Exhaust (At Rest) Position

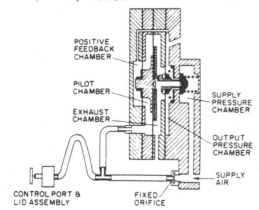

B: Intermediate Position (Exhaust Valve Closed)

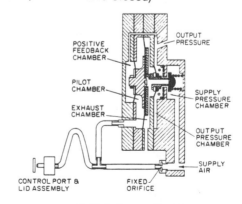

C: Full Output Position

2-13. Pneumatic Off/On, Two Position Sensor

metered to the room. However, the outdoor temperature sensor cannot determine whether the heat input is satisfactory, inadequate or excessive. This lack of feedback defines this as an open loop system.

Figure 2-14b shows a **closed loop system**. Here, the room controller Rt measures and responds to temperature changes by activating the control device providing heat input. Rt also responds to changes initiated by Hi. This feedback closes the loop, making the components interdependent.

Despite their limitations, open loop systems are applicable when precise control is not essential and cost reduction is a factor. However, the closed loop system is more acceptable for maintaining comfortable interior environments.

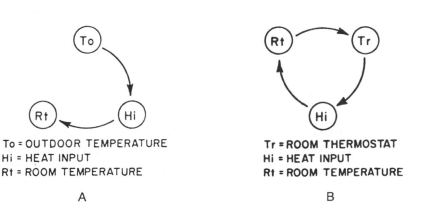

To = OUTDOOR TEMPERATURE
Hi = HEAT INPUT
Rt = ROOM TEMPERATURE

A

Tr = ROOM THERMOSTAT
Hi = HEAT INPUT
Rt = ROOM TEMPERATURE

B

2-14. Open and Closed Loop Temperature Sensing

DDC Proportional Control Systems

Solid state electronic proportional systems were regarded as the ultimate in control sophistication at one time. However, this has changed because modern technology requires that all and even seemingly insignificant variables be factored into all formulas. Today's control systems demand absolute precision—anything less can be disastrous since automation systems also control fire alarms, security functions, and other measures for occupant safety.

In the past, not all variables in proportional control were considered. However, this was not an oversight by control designers. At the time, no logic device existed that could evaluate all input variables required for faultless performance. And since environmental control technology is less strict than modern technology, a degree of inaccuracy didn't constitute a problem.

The advent of computer technology was a turning point in hvac/r. Computers allow sensed inputs to be fed to the computer's arithmetic logic unit (ALU). The ALU then evaluates these inputs based on a mathematical formula that factors in all of the variables required for precise solutions. Such accuracy was not previously possible. In fact, some inaccuracy was expected and necessarily tolerated.

Before computer technology, proportional systems tolerated a certain amount of error termed **offset, drift** or **droop**. Droop occurs, for example, when a proportional hot water valve is 50% open, room temperature controller is at set point 70° F and an increasing heat load creates an error signal. This deviation causes the valve to open and increase the flow of hot water to meet the demand requirement. However it never returns to its original temperature under the existing conditions because an inherent offset error (Figure 2-3) is required to maintain the valve in position to meet the increased demand. Consequently, due to changing demand requirements, the conventional proportional system never maintains zero set point. What is actually sustained is a control point which is equal to the set point plus or minus droop.

Although droop cannot be eliminated, it can be minimized by narrowing the proportional band or throttling range of the controller. Keep in mind that an extremely narrow proportional band causes the system to oscillate and hunt. This short cycling can be more troublesome than the offset error itself.

Offset error is not a serious problem in proportional systems when load changes are slow and slight. However, offset error plus thermal lags and leads and equipment inertia can create variations in control point. In control circuitry, **lags** result from sluggishness in the mechanically control devices caused by hysteresis and inertia.

Figure 2-15 illustrates the lags which can occur in a refrigeration application. When load changes are slight, the resulting changes in control points are tolerable. However, with sizeable and fluctuating loads the delayed remedial response creates wide temperature swings.

Lag #1 can be significant on system start-up but is not usually a major factor in a fully operational 24-hour refrigeration application. It may also be significant in institutional air conditioning applications, where energy management programs dictate decreased or discontinued operation during off-peak loads.

Lag #2 is significant in most applications, as thermal response is not immediate even with large and elaborate equipment. The ability to handle a demand load is the function of system capacity and time, not size or form.

Lag #3 can be reduced but not eliminated with improved solid state sensors; this also applies to lag #4 where solid state circuitry responds with the speed of light.

Changing load requirements, control offset error and system lags often cause faulty system performance. Institutional air conditioning systems are especially vulnerable to these complications. Human psychology also tends to compound the problem. Many individuals believe they are too hot or too cold, despite the fact that instruments accurately document air qualities well within ASHRAE's comfort zone psychrometric parameters.

Proportional Band Plus Reset

Figure 2-16 shows proportional control with inherent offset error. If offset error is eliminated or minimized, total system performance improves substantially. Prior to DDC technology, a manual reset method was used to shift the controller's

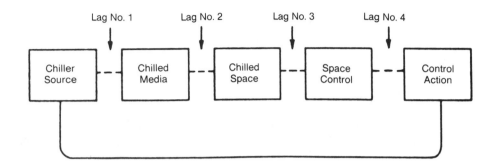

2-15. Environmental System Lags

proportional band upward or downward. This motion opened and closed the valve or control device according to set point requirements.

Provided stable load conditions exist, the manual method adequately minimizes offset error. The manual method requires skilled control technicians to determine the best reset setting after a trial and error period. However, because major load changes can counteract the efforts of even the best control technicians, DDC technology has devised better ways to eliminate offset error.

PROPORTIONAL PLUS INTEGRAL (PI)

A better solution to offset error involves a constant monitoring method that initiates immediate reset to meet existing load conditions. The final controlled element or control device moves at a speed equal to the deviation of the controlled variable.

This corrective action can be mathematically calculated when all the variables involved are factored into the reset formula. The formula must consider two important values:

1) the magnitude of the error signal (which is the proportional variable); and

2) the time integral of the error signal (which is the magnitude of the error signal multiplied by the length of time it has persisted). This is the variable which gives this correction method the name proportional plus integral.

The formula for integral control (automatic reset) is:

$dy/dt = -K(V-Vo)$

where:

dy/dt = Final control element times the rate of motion

V = Controlled variable value

Vo = Controlled variable initial value

K = Constant

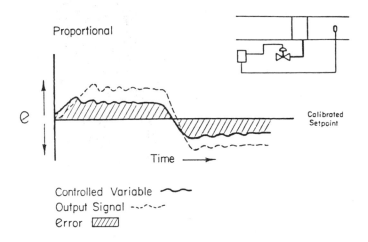

Controlled Variable ⌒⌒
Output Signal ---`--
Error ⊿⊿⊿

Output signal proportional to error in controlled variable.

Adjustment—gain or sensitivity (linear relationship between value of controlled variable and final output signal).

System stability and deviation from setpoint (drift or offset) depend on gain adjustment.

2-16. Proportional Control System Operating Characteristics
Courtesy of Johnson Controls, Inc.

45

A microprocessor can be programmed with a reset formula which begins corrective action as soon as input signals indicate an offset error. In effect, the error signal positions the valve or control device in proportion to the signal. If the offset error persists, the microprocessor calculates the integral value and shifts the output of the control device. This causes the controlled variable to return to the set point temperature. Figure 2-17 shows how system error, integrated over time, results in a zero set point.

DERIVATIVE ACTION

Although integral control (or automatic reset) added to proportional control is the answer to inherent offset (droop) on proportional control systems, it also has a limitation. If the rate of reset is too fast, control system hunting and cycling often

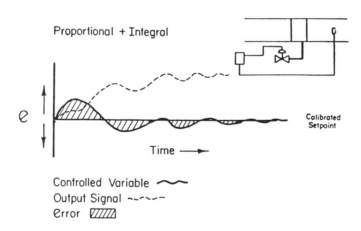

Output signal varies with the magnitude of the error plus the length of time the error in controlled variable occurs.

Adjustments—proportional gain plus integral gain (repeat of proportional signal/time) ratio of the rate of change of the output signal to the error of the controlled variable.

System stability depends on proportional gain and integral gain (time).

Integral control reduces drift but often increases instability which can produce overshoot on sudden upsets.

2-17. Proportional + Integral Control System
Courtesy of Johnson Controls, Inc.

result. If the reset rate is too slow, deviation or drift from set point can occur.

An added control refinement called **derivative action,** rate action or rate response overcomes this limitation. It considers another very important control variable: the error signal's rate of change. Although integral control accurately correlates the magnitude and time of the error signal, it cannot anticipate a rapidly or slowly changing error signal. This inability can lead to hunting or drift.

Derivative action, on the other hand, senses the rate of a very rapid or very slow load change. It also initiates much better corrections than does integral action. Derivative action resembles a professional marksman who can gage the speed of a rapidly moving target and aim ahead of it for optimum accuracy. In contrast, integral action, the amateur marksman, would be aware of the moving target but could not anticipate the rapid rate of change. Integral action also cannot anticipate a very slow rate of change. This lack of anticipation capability delays the return to set point. Derivative action's ability to anticipate additional responses required by the control device provides a prompt return to set point.

The formula for derivative action is as follows:

$(Y-Yo) = -K(dv/dt)$

where:

Y = Final controlled element position

Yo = Final controlled element initial position

K = Constant

dv/dt = Time rate of variable change

PROPORTIONAL PLUS INTEGRAL & DERIVATIVE (PID)

Proportional control is capable of handling the magnitude of the error signal. Integral control is an added refinement that eliminates offset error by multiplying the magnitude of the error by its persistence time. Derivative control handles the time rate

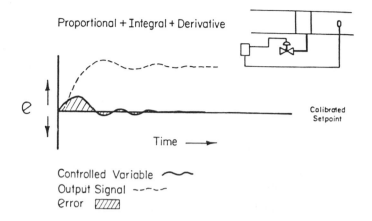

Proportional + Integral + Derivative

Calibrated Setpoint

Time ——▶

Controlled Variable ⌒⌒
Output Signal ⌐⌐⌐⌐
Error ▨▨▨

Output signal varies with the magnitude of the error **plus** the length of time the error occurs **plus** the rate of change of error in the controlled variable.

Adjustment—proportional gain plus integral gain or time plus derivative gain—a lead adjustment which produces an output preceding the proportional output based on rate of change in error signal.

System stability depends on proportional gain, integral gain (time) and the lead adjustment for derivative.

Derivative control reduces the upsets due to sudden load changes which rarely occur in HVAC control.

2-18. Proportional + Integral + Derivative Control System
Courtesy of Johnson Controls, Inc.

of change and applies corrective action by anticipating the additional response required by the control device. A control system which incorporates all of these refinements, with adjustments to meet the application, provides the ultimate in precision performance. This system is called **proportional plus integral and derivative (PID)**.

Figure 2-18 shows the operating characteristics of a PID system. PID is easily attained in a DDC microprocessor-based control system. All required sensor readings are fed to the microprocessor, which performs the mathematical processes necessary for the correct input. This translates into precision performance unheard of only a few decades ago.

■ SUMMARY

There are six basic control systems: on/off, floating control, proportional control, proportional plus integral, proportional plus derivative, and proportional plus integral and derivative. These systems occupy a range from relatively simple to quite sophisticated, yet each system has unique advantages and disadvantages. As a result, control system designers should be familiar with these unique abilities and limitations in order to find the most appropriate and cost effective system for a given application.

INTRODUCTION TO
AUTOMATED CONTROL SYSTEMS
MICROPROCESSOR
FUNDAMENTALS

Control technology has evolved from manual operation to instrument on/off, floating and proportional control operation. Each step in the evolutionary process represents an improvement over the preceding one. Consequently, greater control and precision is attained with less and less human intervention. The ultimate step in this evolution is total system automation.

Automation is defined as the process of completely automatic operation. An automated system is self-controlling in the achievement of its assigned functions. Human intervention is required only when initiating program changes or correcting malfunctions.

Automation in the hvac/r industry is possible because of technological developments in the computer. Its ability to sense any number of variables and direct them into appropriate action indicates that a system can be completely automated regardless of size or complexity. The host computer can even be located in a remote geographic area.

The energy crisis of the mid-70's created a climate for automated control systems. The need for more efficient energy utilization created a new discipline of hvac/r specialization called **energy management**. Since the industry consumes a great deal of energy, it is only natural for hvac/r computers and programs to incorporate the scientific energy-saving techniques developed by formalized energy management studies.

Energy management systems must be aware of all plant or building loads, including lighting and accessory equipment other than hvac/r as well as the related areas of fire alarm, fire safety, and plant and building security. These added variables must be coordinated into the hvac/r program and are now the responsibility of hvac/r system designers.

■ MICROPROCESSOR DEVELOPMENT

Although the computer made total system automation possible, its initial cost proved to be prohibitive for most applications. Only major high-rise buildings, large institutions and industrial plants could justify the high cost of computer controlled operation. Other hindrances to the early development of building automation included hvac/r control industry inexperience and software limitations. Although early systems could monitor potential operating problems, they were not capable of correcting them.

The major development of computer technology was stimulated by the demands of military organizations and the space program. These institutions required smaller computer modules with greater capabilities. Microelectronics provided the miniaturization and in 1970 the microcomputer became a reality. Its features include: smaller size, fewer components, low power needs and greater capability and versatility than the early computers with hard-wired logic.

The early microprocessor was defined as a semiconductor **central processing unit (CPU)** whose elements were contained in a single chip or integrated circuit. The integrated circuit contained thousands of digital gates to perform arithmetic/logic functions and computer operations. CPU's initially required external units such as input/output (I/O) modules and memory, timers, counters and drivers. These units are now contained in the chip itself.

An important factor in the popularity of microcomputer-based automation is its ability to control a single zone in a large complex with a small, self-contained, inexpensive **stand-alone control unit (SCU)**. The SCU performs complete control functions required in a process or a zone and also uses energy management techniques for economical operation. SCU's may also be coordinated in operation with units from other networks or zones in a system. This coordinated operation known as **networking** allows all SCU's to communicate with each other and to be directed in energy management functions by a central computer. Moreover, if a sensor is common to a group of SCU's (e.g., an outside air temperature sensor), only one is required for the entire network.

Of greater importance is the fact that the malfunction of one SCU does not affect the operation of other network units. Even loss of power in the entire system does not affect the network because each SCU is equipped with battery back-up. Enough back-up power is provided to cover nearly any power loss situation.

In practice, an SCU employs features and refinements that give it the capacity and abilities of a microcomputer. However hardware in everyday use consists of a true microcomputer incorporating a CPU or microprocessor which receives, interprets and processes incoming data, and provides an accurate output response. Consequently, it is important to know its structure and how the microcomputer performs its processing functions.

■ LOGIC & LOGIC GATES

Logic involves decision-making based on analysis of incoming signals. All systems (mechanical, pneumatic, fluid control systems and solid state signal centers) perform logic functions. For example, a single-pole, single-throw electrical

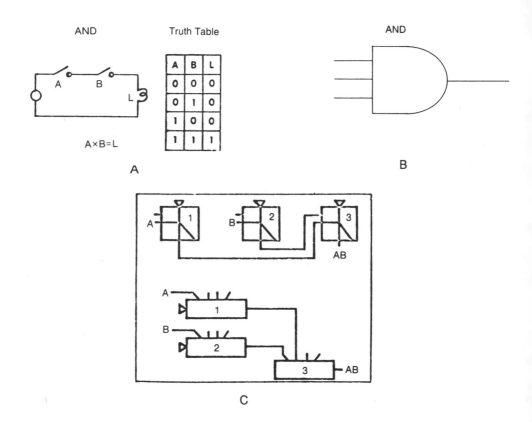

3-1. AND Gate Logic Circuit and Symbol

switch performs a logic function through mechanical action. When a temperature sensor signals the closure of the switch, it makes a logical decision to energize an electrical circuit which closes relays and operates the compressor.

However, sophisticated microcomputer functions require the use of digital logic gates. A **logic gate** is a circuit whose output is de-energized until certain input conditions have been met. Each of the four basic gates (AND, NAND, OR and NOR) has individual configurations and input conditions.

AND GATE

If the previous example required two simultaneous events to operate the compressor, the logic function would become more complex. As Figure 3-1 shows, this could involve two single-pole, single-throw electrical switches in series. If switch A is open and B closed, the electrical circuit is not completed and output device L is not energized. The same condition exists if B is closed and A is open.

However, if switches A and B are both closed, the circuit is complete and output device L is energized. The truth table tabulates what the action of the output device will be when various input modes to the switches are in effect. In this case, a refrigeration compressor can start only if a temperature sensor closes one switch and if a low pressure control in series keeps the other one closed. In the language of logic circuits, this is the well-known **AND** gate. The AND gate functions the same whether the circuit is mechanical, pneumatic, fluidic or solid state. It can be symbolized graphically (Figure 3-1a) or mathematically (for formula computations).

Figure 3-1b shows the electronic version of the AND gate. In the AND gate symbol, the signal always flows from the flat side to the round side. The round side always represents the output. Figure 3-1c depicts a fluidic AND gate. Signal flow is always from left to right.

In the AND gate, no output occurs if:

- there is no input signal
- there is a signal at A only
- there is a signal at B only.

However, if there is a signal at both A and B, there is an output at AB.

In short, the AND gate only delivers output when it senses two incoming input signals.

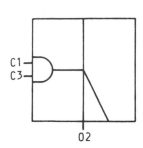

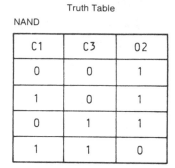

Truth Table

NAND

C1	C3	02
0	0	1
1	0	1
0	1	1
1	1	0

A

NAND

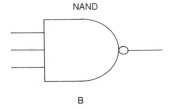

B

3-2. NAND Gate Logic Circuit and Symbol

NAND GATE

Figure 3-2 illustrates the **NAND** gate. This gate (as the truth table verifies) is the exact opposite of the AND gate. Output occurs if there is no input at C1 and C3 or if there is input at either C1 or C3. Figure 3-2b depicts the electronic symbol for NAND. In effect, the NAND gate is an inverted AND gate.

OR GATE

Figure 3-3 is an example of the **OR** logic gate (Figure 3-3a is an electromechanical version of an OR gate). Basically, this configuration is the same as two single-pole, single-throw switches in parallel connection to a load device. The load device is energized if either switch is on or if both are on. Note that in the electronic symbol for OR (Figure 3-3b), the signal always runs from the concave to the convex side. The convex side always represents the output.

Logic gates can have variations in logic, as Figure 3-3c illustrates. This variation is called **Exclusive OR**. It differs from conventional (non-exclusive) OR in that there is no output if both inputs are on. This logic is difficult to perform electromechanically but is easily conducted with transistors or fluidic devices. Figure 3-3c illustrates the fluidic symbol for OR.

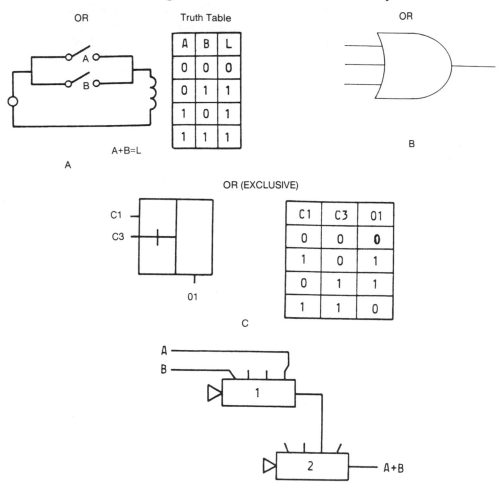

OR   Truth Table   OR

A	B	L
0	0	0
0	1	1
1	0	1
1	1	1

A+B=L

A   B

OR (EXCLUSIVE)

C1	C3	01
0	0	0
1	0	1
0	1	1
1	1	0

C

A+B

3-3. **OR and EXCLUSIVE OR Symbols and Truth Tables**

57

NOR GATE

The **NOR** gate (Figure 3-4a) has output only when both inputs are off. **Exclusive NOR** differs from NOR in that output occurs if both inputs are on *or* off. NOR and Exclusive NOR have no output if there is input at only C1 or C3.

Figure 3-4b is the **NOT** symbol, which performs an inverting function. It has output with no signal and it turns off if there is an input signal. The electronic symbol for NOR is shown in Figure 3-4c.

The AND, NAND, OR and NOR gates represent the basic logic devices that form the building blocks of complete logic circuits. These logic gates are depicted by symbols. The sym-

NOR

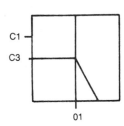

01

A

C1	C3	01
0	0	1
1	0	0
0	1	0
1	1	0

NOT

02

B

C1	02
0	1
1	0

NOR

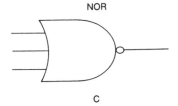

c

3-4. NOR Symbol and Truth Table

bols represent logic functions performed by transistors, diodes and resistors (in solid state operations) or by fluidic devices (pneumatically). However, computers only utilize electronic or solid state logic devices.

■ COMBINATIONAL AND SEQUENTIAL CIRCUITS

Digital control circuitry is classified into two basic types: combinational and sequential. In a **combinational circuit,** the output is determined by the action of the existing inputs. The preceding logic gates are suitable for combination digital control circuitry. Combinational circuits have no memory and are not dependent on the past sequence of events.

Sequential circuits, on the other hand, contain logic and memory elements. These store information contained in the past sequence of events with the help of the flip-flop logic device.

FLIP-FLOP

The memory elements of a sequential circuit are contained in the **flip-flop logic device** (Figure 3-5). It possesses memory because the circuit retains output conditions as a function of the last input. It remembers and maintains this output even if the input signal is removed.

The truth table indicates that with no signal input at A and B, no output occurs at F1 but an output does occur at F2. In the second sequence of events, an input at A creates an output at F2 and no output at F1. In the third sequence, an input at B results in output at F1. In the last sequence, input at A and B results in output at F1.

The preceding flip-flop truth table illustrates the reset and set (R-S) flip-flop mode. Flip-flop logic devices provide bi-stable operation for the memory functions demanded by computer operation.

Other flip-flop forms include the toggle (or complementary) flip-flop and the J-K flip-flop. The toggle has only one input

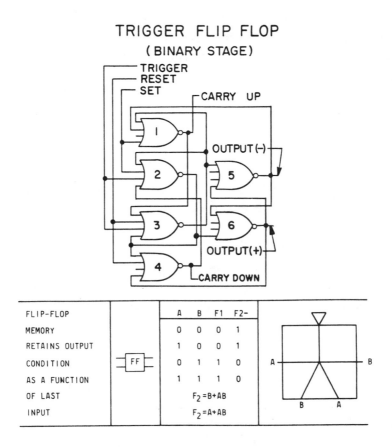

3-5. Flip-Flop Symbol and Truth Table

terminal. Each signal at this input terminal sends the flip-flop into its opposite state. Thus it goes from set to reset in a manner similar to a mechanical toggle switch in an electrical circuit. The J-K flip-flop has the greatest usage in microcomputers. It also has two inputs similar to the R-S flip-flop, but the J-K flip-flop can have both inputs activated simultaneously and it will change state regardless of its previous mode.

It is not essential that hvac/r professionals have a complete and specific knowledge of all microprocessor chips and circuits. However, an awareness of its capabilities and limitations is invaluable in microprocessor-based control system design and application.

■ PC COMPONENTS

The PC is a microprocessor-based device. The computer's microprocessor or CPU often consists of a single chip and contains a permanent operating program (firmware) that converts relay logic diagrams into logical instructions which the PC can understand and process. The controlled elements of the CPU can also be contained in a single chip or control panel (although it can involve several chips).

Computer data is processed in binary units called bits. A **bit** is the smallest addressable information unit in computer memory. A group of eight bits is called a **byte.**

An analysis of the major functions in microcomputer operations provides a perspective which may make design, application, maintenance and troubleshooting of a control system more effective. The main components or functions of a microcomputer are:

1. The Central Processing Unit (CPU)
2. Memory
   Read Only Memory (ROM)
   Random Access Memory (RAM)
3. Input/Output
   Analog and Digital I/O
   Parallel and Serial I/O
4. The Program (software)

1. CENTRAL PROCESSING UNIT (CPU) COMPONENTS

The CPU or microprocessor is the core or "brain" of the computer. It is the CPU which coordinates the actions and information of all the other components such as input/output, memory and program instructions.

Figure 3-6 illustrates the memory structure of the CPU itself. This structure involves the input memory unit, program memory, variable data memory and output memory.

An **input memory unit** provides a memory location for all inputs in the I/O component. It informs the CPU as to the exact input conditions demanded by the sensors. Digital or binary language conveys this as a binary 1 for 'on' (HI) or a binary 0 for 'off' (LO).

The **program memory unit** is where the programmed instructions are stored. During programming, the operator uses a ladder relay logic wiring diagram to encode and link the diagram wiring sequence and symbols with the PC manufacturer's CPU symbols. The operator then manually selects the CPU's program mode and enters the coded instructions sequentially into the program memory unit. The number of program instructions can range from less than ten to thousands, depending upon the program requirements and the PC's storage capabilities.

After programming, the PC is placed in the operating mode. This mode initiates the repetitive execution of the program according to prescribed instructions. The numerical information necessary for system operation is contained in the

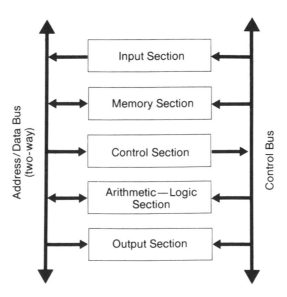

3-6. A Simple Microprocessor System Configuration

CPU's **variable data memory**. As new values are needed and calculated, they are stored in this memory section.

The **output memory table** has a memory location for each individual output circuit. It also monitors the status of each output device. Specific locations in memory are called **addresses**, and each address has its own specific binary code (however, the actual coding depends upon the manufacturer). The output memory table feeds instructions or information to the output module and on to the output device. It can also receive feedback information from the output device and return it to the Central Processing Unit.

The CPU executes a program by coordinating the input signals (from the input memory table), information data (from the variable data memory) and output information (from the output memory table) with instructions from program memory.

To process incoming data, the CPU draws upon other internal resources besides its memory units. The CPU also contains these interconnected functional blocks:

- Register and Address Section
- Arithmetic/Logic Section
- Instruction Register and Control Section

Register and Address Section

This section contains the registers for temporary data storage within the CPU. The data may be general or specific. The general–purpose registers are often designated by a single letter. In contrast, registers which have dedicated (specific) uses include the program counter, stack pointer and instruction register.

The **program counter** is one of the registers holding addresses necessary to lead the microprocessor through various programs. The program counter also holds the address of the next instruction byte to be fetched from memory and is automatically incremented after each fetch cycle. During interruptions the program counter saves the address of the instruction.

Another register component is the **stack pointer** (SP). Stack pointers are coordinated with the storing and retrieval of information in the stack, which is a reserved area of memory.

The instruction register is also part of the register and address section. It holds the instructions being executed by the CPU after being brought to the control section from memory. The instructions allow the microcomputer to perform data manipulations and arithmetic operations.

Arithmetic/Logic Section (ALU)

The ALU section actually processes the information or data and performs any arithmetic and logic operations required by the program. These are the Boolean operations executed

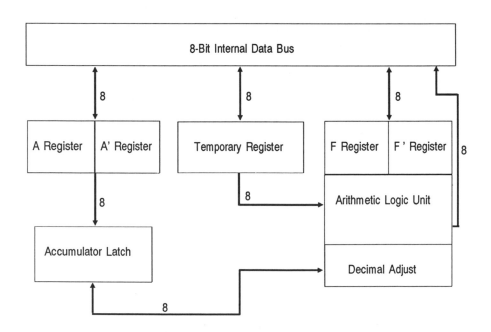

3-7. Internal Section of an ALU Unit
Courtesy of Robert A. Greyell, North Seattle Community College

by the basic logic gates in combination configurations with flip-flop devices and in conformance with the algorithms of the program designer.

This section of the CPU also performs arithmetic and logic operations on the stored binary file. It is composed of the ALU plus an accumulator or 'A' register, temporary holding registers and any necessary flag registers (a flag is an information bit which indicates a form of demarcation). The ALU also contains a circuit that can add the binary contents of the accumulator and any other register. This provides a method for arithmetic

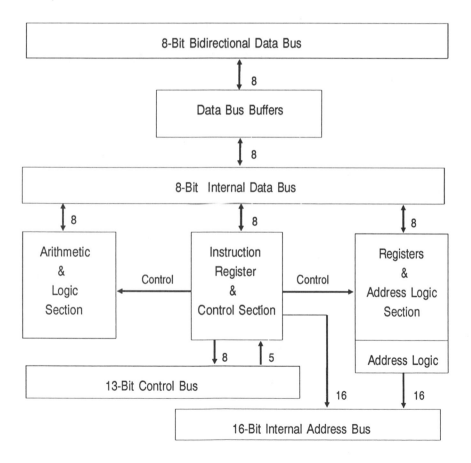

3-8. Internal Configuration of a CPU
Courtesy of Robert A.Greyell, North Seattle Community College

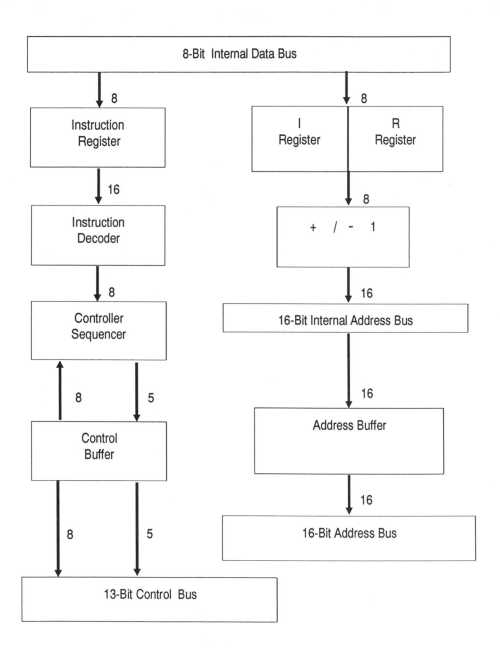

3-9. Functional Configuration of a CPU
Courtesy of Robert A. Greyell, North Seattle Community College

computations on data either stored in memory or obtained from the input section.

Figure 3-7 is an internal view of the ALU section shown in Figure 3-8. It contains two independent accumulators, A and A', and two Flag register pairs, F and F'. The accumulator receives the result of all eight-bit arithmetical and logical operation. On the other hand, the Flag (F) register indicates the occurrence of specific logic or arithmetic conditions. The bits within the selected Flag register specify the CPU conditions which have occurred after arithmetic, logic or other specific CPU operations.

Additional functions that may be built into the ALU are hardware subtraction, Boolean logic operation, data shift operation and the ability to detect data meeting specified conditions. In addition, ALU's may have a hardware multiply-and-divide function to provide high speed mathematical execution.

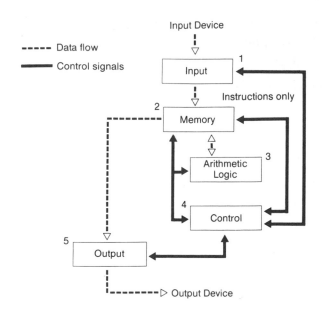

3-10. Basic Microprocessor Configuration
Courtesy of Robert A. Greyell, North Seattle Community College

Instruction Register And Control Section

This is the primary functional unit in the CPU (Figure 3-9). The control circuitry decodes the incoming instructions, synchronizes all microcomputer action and insures that each data processing task functions in proper sequence. Figure 3-10 shows how the CPU utilizes the control section and the ALU to carry out a program. The control section provides a flow of information between the memory and the ALU.

The control section may also be designed to respond to external signals, such as Interrupt or Wait requests. The Interrupt request temporarily stops execution of the main program, services the interrupting I/O device and then returns to continue the main program. A Wait request idles the CPU until data is ready. The Wait request can be initiated by I/O devices, memory elements or both because these devices and elements operate at a much slower rate than the CPU.

2. MEMORY

The **memory** is the storage area for data and instructions essential to arithmetic and logic operations. It also provides information on output and control of operations.

The main memory for program and data storage is usually in the form of 'firmware' or permanently wired memory. This uses interchangeable pre-programmed **Read Only Memory (ROM)**. ROM contains the permanent programmable logic, which in hvac/r systems provides the basic operations needed for equipment power-up routines and operating system initialization. Program instructions which do not change are permanently stored in ROM and cannot be erased by power failure. ROM has addressable storage locations and provides access to permanent storage information that is repeatedly used such as complete mathematical tables, predetermined messages or sequence bits required for data communication.

EPROM is Erasable Programmable Read Only Memory. Like ROM, EPROM can store programs and control instructions and is also not erased with power outage. However,

EPROM can be erased when exposed to ultra-violet light. As a result, old programs can be erased and new ones entered as required.

Easily changed data in the form of software is **Random Access Memory (RAM)**. RAM stores variable data or data that may change. Consequently, information and instructions that may change as the hvac/r application changes can reside in RAM. RAM is dynamic and can lose its stored information with power outage. As a result, a battery back-up contingency is usually provided to prevent memory loss from periodic power outage.

In RAM each storage location is addressable. Every word in memory has its own unique address, and a word may be defined as a set of bits. Slower and less expensive memory systems do exist. These are used for long term bulk storage of programs and include paper tape, magnetic tape, punched cards, floppy disks and hard disks.

The most common of the memory elements is the flip-flop. FF's (flip-flops) can be connected to form **registers** when many bits of information need to be stored. However, when many groups need to be stored, FF registers tend to be too physically large and extensive. RAM packages can supply the necessary additional storage. This provides adequate storage for binary data that may only be referenced occasionally. RAM storage may or may not be FF devices.

3. INPUT & OUTPUT

Input/output refers to all the ways in which a computer receives and transmits data. In hvac/r systems the control signal, in the form of a DC voltage output, is received by the input section of the I/O device. The input device can involve digitizers, an ADC, keyboards, card or tape readers, switches and control sensor inputs. These data and instructions are then supplied to the microcomputer's CPU to be processed (Figure 3-11). The CPU reads and analyzes this incoming data and provides a programmed output response. Output devices such as proportional motors, relays, pneumatic actuators, switches, electronic-to-pneumatic or fluidic transducers can be used to initiate, modulate or terminate action by the controlled device.

The control sensors that monitor system performance use line voltage (120V AC) to send their control signals to the input terminal of the I/O device. The input module converts this AC voltage signal into a digital 1 (or 'on') for processing by the CPU. The output module then converts the processed signal by amplification into a line voltage or low voltage output response that initiates action by controlled devices. Conversely, if the input terminal does not receive a sensor signal, it responds with a digital 0 (or 'off'), indicating that no output response is required. In this fashion hvac/r equipment cycles on and off to maintain design parameters.

The preceding I/O description is an example of digital operation. It is typical of sensors that provide on/off signals which in turn cycle motors or motor starters, solenoids, dampers and valves in the full on or full off mode. I/O can also involve keyboard interfacing and data communication between digital devices. In addition to line voltage input and output, the I/O's can interface with other electrical sources such as 5V DC and 24V DC TTL (Transistor-Transistor Logic).

Digital I/O

Digital represents a distinct value which can be yes or no, present or absent, on or off. Consequently, digital adapts to Boolean binary algebra equations which the microprocessor can translate into numerical values. This capability allows the computer to perform an infinite number of functions. Digital is capable of absolute accuracy because it transforms information into precise bits which can be mathematically processed and stored. In this precise form, digital data can be instantly checked for errors. In addition, digital data can be accurately transmitted and received over great distances via communication networks.

Analog I/O

Many operations in modern control systems require more sophistication than on/off sensing and cycling can provide. To maintain zero or nearly-zero set points, the control system may have to sense variable signals and provide variable responses. In this circumstance, system sensing and response must include **analog** capabilities.

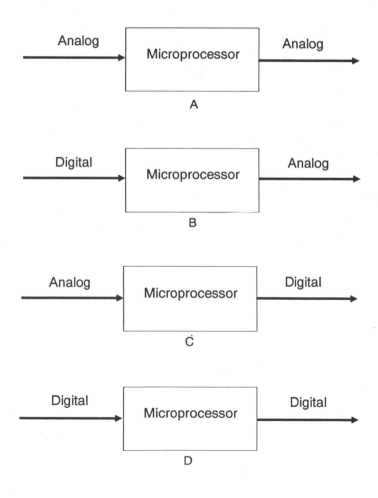

3-11. Microprocessor Input-Output Configurations

71

Unlike digital, analog is a continuous series of controlled variables having a variety of assigned locations. This varying signal constantly measures a controlled variable. The variable can be temperature, pressure, humidity, fluid flow or any value subject to change. As a result, the signal amplitude is in proportion to the stimulus provided by the controlled variable. An analog signal that results in an analog response or output is conducive to proportional control.

Analog input-analog output (Figure 3-11a) is the most utilized method in hvac/r control systems. The analog I/O method lets the microprocessor adapt readily to the analog sensors and output devices which are dominant in the control industry.

However, many configurations require a mix of analog and digital inputs and outputs. Figure 3-11b illustrates digital input/analog output, Figure 3-11c shows analog input/digital output, while Figure 3-11d is digital output/digital input. Since microprocessors are digital, a converter from analog to digital (or vice versa) is necessary.

Peripheral I/O Devices

Input/output devices help provide digital input information to the CPU for processing and digital output instructions for action by output devices.

Proportional control action is dominant in hvac systems because controllers provide analog signals. As a result, output devices (such as proportional motors) must provide an analog response. This requires interface devices to convert the incoming analog signal from the controller into a digital signal which can be processed by the microcomputer.

Since PC's and microcomputers are inherently digital devices, analog sensor signals and output responses must undergo conversion to and from processing. This translation is accomplished by analog-to-digital (ADC) and digital-to-analog (DAC) converters. Thus varying parameters such as temperature, humidity, pressure, voltage, resistance and electrical current can be converted to digital equivalents by the ADC. The digital equivalents are on/off signals expressed as binary units—one or zero (logic high or low). The module is usually multiplexed to handle a number of analog signals on a time-sharing basis.

Figure 3-12 illustrates an analog sensing procedure with a pressure transducer. The transducer shown has a span of 0 to 300 psi. Its output (in the form of a voltage or current signal) is at midpoint when it senses 150 psi. If, for example, output voltage varied from 1 to 5V DC, midpoint pressure would be 3V DC. The 3V DC signal would then be directed to an ADC for conversion to a digital number for the computer to process.

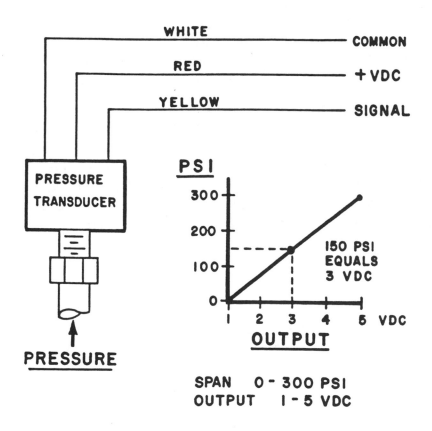

3-12. Analog Sensing with a Pressure Transducer
Courtesy of David A. Murphy, Frank Electric Corporation

The ADC and DAC conversion functions are usually packaged with the I/O hardware.

The absolute precision of the signal depends upon analog resolution. Resolution (signal strength) is specified in bits, and a greater number of bits provides more precise resolution. For example, if the analog resolution shown in Figure 3-12 is twelve bits (4096 increments), the signal accuracy and precision is within one-tenth of one psi in the 0-to-300 psi transducer. With eight-bit resolution (256 increments) signal accuracy significantly decreases to more than 1 psi.

Analog inputs are essential in proportional control. For example, if the transducer in Figure 3-12 required variable response from a proportional or variable-speed motor, the analog signal would be converted to digital for processing by the PC. After processing, the DAC would convert the signal back to an analog output for action by the controlled device.

Other various peripheral devices exist which enhance microcomputer functions. Tape and disc units as well as controllers can be used to input data to the CPU. Video display terminals and even color display CRT graphics provide visual reference in programming, system operation and surveillance. Typewriter keyboards allow the operator total control of microcomputer operation, and printers deliver hard-copy records and reports for documented communication. In addition, a modem—a device which changes signals from analog to digital form—enables computers to communicate over telephone circuits.

A peripheral I/O device called a **bus** also expedites the communication process. The internal arrangement of a typical CPU section containing a bus is shown in Figure 3-8. The eight-bit bi-directional data bus is a conducting path capable of transmitting information to and from the CPU. As shown, this data bus transmits with a width of eight bits (one byte). The data bus connects data from I/O devices, RAM and the CPU registers. In Figure 3-8, thirteen CPU and system control signals can be initiated or received from the instruction decode and CPU control portion of the microprocessor.

The **data bus buffer** expedites data flow by supplying additional drive. It can also delay the rate of information flow for more effective processing. The internal eight-bit data bus is the main conducting path for information within the CPU. It connects the CPU register, the ALU, the data bus control and the instruction register. The sixteen-bit address bus allows direct addressing for up to 64K (65536) locations.

In addition, a **clock function** must be incorporated in a microcomputer system in order to provide timing needed to synchronize the functions of ALU, ROM and RAM units. An **address selector** provides a selection method to determine which circuit an address will access, and the **driver circuit** is

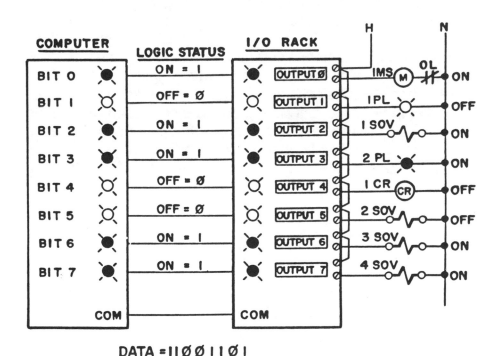

3-13. Parallel Addressed Digital Input-Output Device
Courtesy of David A. Murphy, Frank Electric Corporation

an element coupled to the output stage of a circuit to increase its load handling capabilities.

Parallel and Serial I/O

There are two types of I/O data transmission, parallel and serial. They are based on the method by which the output line is turned on and off in an I/O device.

Parallel I/O is a digital form where more than one output line (bit) is turned on or off at one time (illustrated in Figure 3-13). Each input or output requires its own conductor. Parallel I/O ports are eight bits wide or in multiples of eight bits wide. Individual bits within a byte-wide port can be turned on or off through the use of bit masks which are selected to fit only the desired bit.

Parallel I/O is also used for data communications. In Parallel I/O each bit has its own conductor but multiple bytes are transmitted serially. This arrangement is called **bit parallel/byte serial**. This simultaneous communication of multiple bits gives parallel interfacing a slight speed advantage over serial, but this is of no great significance in most industrial applications. Parallel I/O is used to switch individual controlled outputs on and off, and to read the status of on/off type input switches. It can also be used for keyboard inputs and display screen outputs, usually in conjunction with some type of multiplexing scheme.

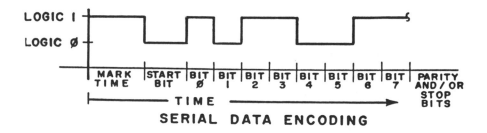

SERIAL DATA ENCODING

3-14. Digital Serial Input-Output Device
Courtesy of Frank A. Murphy, Frank Electric Corporation

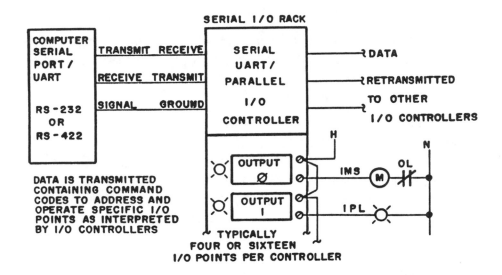

SERIAL I/O RACK

3-15. Serial Data Transmission
Courtesy of David A. Murphy, Frank Electric Corporation

Parallel transmission has the advantage of being faster than serial I/O, but parallel interfacing has hardware limitations (such as short cable lengths) which require the I/O device to be quite close to the microcomputer, which limits the use of remote peripherals. Also, parallel interfacing has a limited number of addressable I/O devices due to the limited number of parallel ports.

In contrast, **serial I/O** (another form of digital) sends the multiple bits of information over the same output path, one bit at a time (Figure 3-14). Serial I/O is generally grouped in a succession of bytes. A few control bits are often added to each byte for parity checking or signaling the end of the byte (i.e., parity and stop bits).

The speed of serial I/O is measured by the baud rate. The **baud rate** equals the number of data bits transmitted per second. The common baud rates for asynchronous serial communication are 300, 600, 1200, 2400, 4800, 9600 and 19,200. The baud rate divided by 10 gives an approximation of the number of bits that can be transmitted per second.

A disadvantage of serial I/O is that it is slower than parallel I/O. However, serial I/O has the advantage of operating at a greater distance from the PC. The actual distance depends upon the specific hardware involved.

Most data is transferred serially because it can be transferred over great distances and over a common set of conductors (instead of individual conductors as in parallel I/O). As a result, serial I/O can be used for mass data transfer between solid state devices such as system controllers, host computers, printers and telephone modems. Serial I/O is also used to address groups of remote I/O devices through serial communication modules resident on the I/O rack assembly. This arrangement is ideal for interfacing with master control computers because the total number of I/O ports is not limited by parallel port availability.

Because it is inconvenient to communicate digital data over long distance in a parallel mode, data can be transmitted

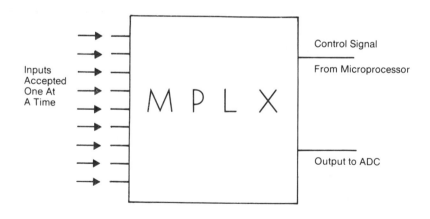

3-16. Multiplexer Transmission Switching Component
Courtesy of Johnson Controls, Inc.

in a serial format and then converted back to parallel format for entry on the microcomputer data bus. A specialized circuit module called a **Universal Asynchronous Receiver Transmitter (UART)** encodes and decodes the data to accomplish the conversion. A UART converts the data to a serial bit stream at the measurement site. Then it is sent on a single conductor to the microcomputer where another UART converts it back to individual parallel bits (Figure 3-15).

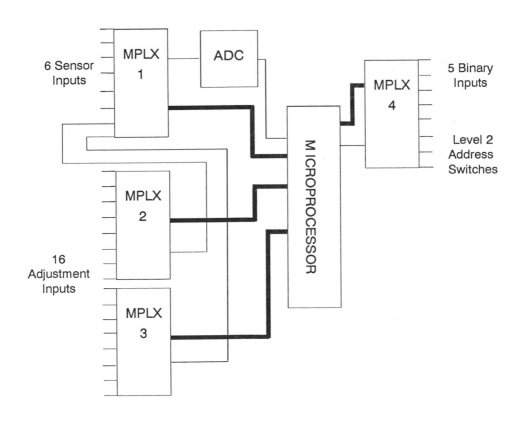

3-17. CPU Using Four Multiplexer Transmission Units

Courtesy of Johnson Controls, Inc.

Another transmission component is the **multiplexer** (MPLX), shown in Figure 3-16. This is an electronic multiposition switch controlled by the CPU and used in conjunction with an ADC. It is directly connected to the CPU and can be digitally addressed to select and channel one of a number of analog signals as input to the ADC. The signal is then sent by the ADC to the CPU for processing according to software instructions.

The CPU configuration in Figure 3-17 uses four eight-input MPLX's (a,b,c and d). MPLX #1 has six inputs for sensors and two inputs to the output ports of #2 and #3. MPLX #4 accepts binary information and therefore does not require a connection to port #1.

4. THE PROGRAM

Although the CPU is the "brain" of the microcomputer, it needs a **program** to provide it with function requests. Programs (software) contain the instructions which direct the CPU in a process. (Programming is addressed more thoroughly in the following sections.)

■ SUMMARY

The evolution of the microcomputer launched a simultaneous evolution in hvac/r system automation. Obvious advantages of the computer include more efficient operation, as well as greater speed and accuracy.

Although a more detailed analysis of microcomputer internal functioning and system programming may be helpful, it is not essential for microprocessor-based control system design and application. Subsequent sections describe any pertinent specifics in this area.

PROGRAMMABLE CONTROLLERS & PROGRAMMING BASICS

The **programmable controller** (**PC**) is a device that programs a CPU (Central Processing Unit or microprocessor) to direct and monitor conventional controls in industrial hvac/r applications. The programmable controller is capable of automation despite the fact that it was initially developed as a transition from the solid state signal centers to the dedicated (designed for specific use) microcomputer.

The term "programmable" refers to the fact that PC's may be field-programmed by the user. PC programming requires no specialized knowledge of computer assembly language, memory or hardware. The user simply follows the control system's ladder diagram as a ladder-logic format for programming. The PC manufacturers typically provide the instructions and symbols which translate the ladder diagram into language the microprocessor can understand. Consequently, a PC can execute programs involving relays, counters, drum timers, step switches and even the analog functions of pressure and temperature controls. In addition, it can provide math and logic functions and generate messages.

■ PC ADVANTAGES & DISADVANTAGES

Despite its abilities, the programmable controller does not fully possess the versatility and programming capabilities of a dedicated microcomputer. The PC is limited in the number of inputs and outputs it can handle, as these may require many wires while the dedicated microcomputer can require as few as two.

However, the PC does have its advantages. It is quite capable of handling and operating an individual system involving related equipment. Also, the ability to easily understand a PC (because of its ladder logic format) allows greater programming flexibility and modification. Troubleshooting and de-bugging are also simplified. In addition, the PC adapts better to hostile environments (e.g., dust, temperature and humidity extremes) than do dedicated microcomputers. Therefore, the PC has advantages in limited areas despite the overall capabilities of the dedicated microcomputer.

The PC is most effective and least costly if the system to be controlled has a limited number of equipment operations. For example, this could involve the sequencing of two or three screw compressors in parallel, or four to six reciprocating compressors plus several steps of condenser control. However, with additional equipment and control requirements (which add input and output capability) the dedicated microcomputer has an advantage in both cost and operating effectiveness.

In short, the PC is very easy to apply and program in relatively small systems. PC input/output (I/O) interfacing and addressing can be easily accomplished by using manufacturers' instructions. As a result, the relay logic and ladder diagram should present no serious knowledge barrier to the trained hvac/r technician.

■ **PC PROGRAMMING**

Programming is the process of constructing a computer program for a particular application. The first step in developing a program for an hvac/r system is to design an accurate, systematic ladder diagram.

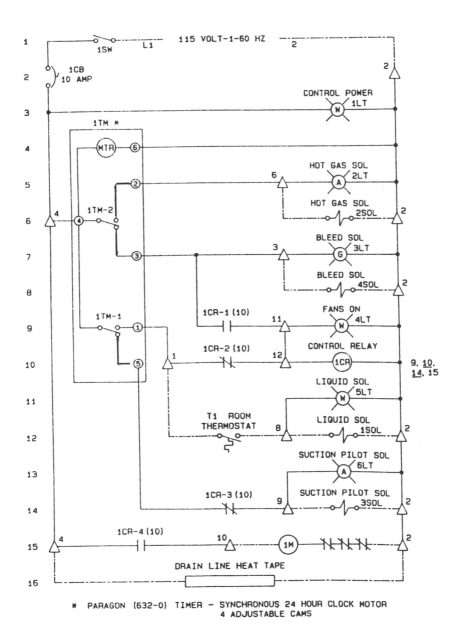

4-1. Ladder Diagram for a Hot Gas Defrost System
Courtesy of Allen-Bradley

LADDER DIAGRAMS

Ladder diagrams are schematic figures which make it easier to visualize and identify a circuit's components and connections. A ladder diagram consists of two vertical parallel lines (representing the main power source) with various devices and their connections shown on the horizontal lines (referred to as rungs) in between.

Figure 4-1 illustrates a basic hot gas defrost system in a ladder-logic format. The left side of the diagram is numbered 1 through 16 (numbering is essential in ladder diagrams in order to establish precise reference parameters). The numbers represent imaginary horizontal lines called rungs or steps because they resemble rungs of a ladder. Just as climbing a ladder involves a rung-by-rung approach, the PC user's program is also scanned sequentially to provide the prescribed system performance.

The typical relay diagram in Figure 4-2a would be a rung in a conventional wiring diagram. Figure 4-2b depicts the same

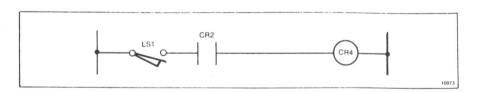

A. Relay Diagram

B. Ladder Diagram Rung

4-2. Conventional and PC Diagram Rungs
Courtesy of Allen-Bradley

diagram using the PC user's program symbols. If depicted on a PC monitor screen (CRT), the foregoing rung would appear similar to that shown in Figure 4-2b.

Figure 4-2a shows a basic circuit to cycle a refrigeration compressor. In ladder wiring diagram rung 1, limit switch LS1 represents the contacts of a high-low pressure control protecting a refrigeration compressor. CR2 is a temperature-sensing operating control and CR4 is a relay pull coil in the motor starter. In actual practice other safety controls (such as lube oil protection control) might be included in the circuit.

The manufacturer's symbols and instruction keys are all shown in a typical industrial terminal (Figure 4-3). The terminal keyboard is displayed on the CRT screen. The PC can then take relay instructions from the ladder electrical diagram or instruction manual. By converting the symbols to the appropriate keys on the terminal keyboard, the PC can provide the following capabilities:

Examine instructions

Output instructions

Branch instructions

Examine Instructions

The two types of **examine instructions** are Examine On | | and Examine Off |/|. These are the same as electrical switch functions.

Examine On symbolizes the equivalent of a normally open electrical contact. When power (in the form of an electrical signal) is applied to the proper terminal in the input module, the processor scans the related memory bit. When it determines this to be input ON (or 1), it verifies that electrical continuity of the circuit has been established to this point in the rung. The Examine On instruction addresses the I/O image table to determine if condition ON exists. If it does exist, this means it is True and an electrical circuit is complete to this point. The Examine

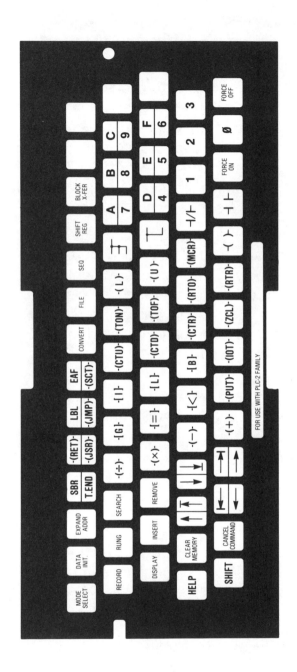

4-3. Programmable Controller Keyboard
Courtesy of Allen-Bradley

On condition can only be True if a circuit is on. It is False if a circuit or voltage is unavailable.

Examine Off is the opposite of Examine On. Examine Off tells the processor to check the I/O image table for an OFF (no voltage) condition. In this mode it can examine a single input or output bit for a no-voltage state. A True condition in this mode indicates that the I/O device (or bit) is off, while False indicates that the I/O device is on. Examine Off resembles a normally closed contact in an electrical diagram but requires modification to be electrically valid in the PC.

Output Instructions

Output instructions in the PC are programmed at the end of ladder diagram rungs. The method used is similar to programming relay logic ladder diagrams. The output instructions also set an addressed memory bit to 1 (ON) or 0 (OFF). Only one output instruction is programmed in each ladder diagram rung. The output instruction is executed only if the preceding instructions in the rung are valid and True.

The output instructions and their PC symbols are:

Output Energize	-( )-
Output Latch	-(L)-
Output Unlatch	-(U)-

The **Output Energize** instruction tells the CPU to turn on an addressed memory bit when the rung conditions are True. In this manner the memory bit determines the status (on or off) of the output controlled device. When rung conditions read False, Output Energize tells the processor to turn off the addressed memory bit.

Output Latch is similar to Output Energize with one exception. Output Latch will instruct the processor to turn on an addressed memory bit when rung conditions are True. In this mode Output Latch is retentive (i.e., if the rung conditions are

RELAY INSTRUCTIONS

NOTE: Examine and Output addresses, XXX/XX, can be assigned to any location in the Data Table, excluding the processor work areas. The word address is displayed above the instruction and the bit number below it. To enter a bit address larger than 5 digits, press the [EXPAND ADDR] key after the instruction key and then enter the bit address. Use a leading zero if necessary.

Keytop Symbol	Instruction Name	1770-T3 Display	Description
–| |–	EXAMINE ON	XXX –| |– XX	When the addressed memory bit is ON, the instruction is TRUE.
–|/|–	EXAMINE OFF	XXX –|/|– XX	When the addressed memory bit is OFF, the instruction is TRUE.
–( )–	OUTPUT ENERGIZE	XXX –( )– XX	When the rung is TRUE, the addressed memory bit is set ON. If the bit controls an output device, that output device will be ON. *(1)*

4-4. Programming Symbols, Terminology & Instructions
Courtesy of Allen-Bradley

-(L)-	OUTPUT LATCH	XXX -(L)- ON XX or OFF	When the rung is TRUE, the addressed memory bit is latched ON and remains ON until it is unlatched. The OUTPUT LATCH instruction is initially OFF when entered, as indicated below the instruction. It can be preset ON be pressing a [1] after entering the bit address. An ON will then be indicated below the instruction in PROGRAM mode. *(1)*
-(U)-	OUTPUT UNLATCH	XXX -(U)- ON XX or OFF	When the rung is TRUE, the addressed bit is unlatched. If the bit controls an output device, that device is deenergized. ON or OFF will appear below the instruction indicating the status of the bit in PROGRAM mode only. *(1)*
┬	BRANCH START		This instruction begins a parallel logic path and is entered at the beginning of each parallel path.
┴	BRANCH END		This instruction ends two or more parallel logic paths and is used with BRANCH START instructions.

(1) These instructions should not be assigned Input Image Table addresses because Input Image Table words are reset each I/O scan.

(4-4. Continued)

False, the latched bit still remains on). It can only be de-energized when the Output Unlatch instruction is given. The **Output Unlatch** instruction is also retentive—when rung conditions are False, the unlatched bit remains off.

Branch Instructions

As electrical circuits are composed of series and parallel branches or a combination of both, the PC must be able to program both. Branch instructions allow the PC to read parallel circuitry.

Series programming is relatively simple, as each input is programmed sequentially with its appropriate instruction followed by an output instruction. On the other hand, parallel circuits must allow for more than one input condition to energize an output device. This is a form of OR logic but the PC must process it through its own symbols, terminology and instructions (Figure 4-4).

Parallel circuitry has two instructions, Branch Start and Branch End. Enter **Branch Start** 1 before the first input instruction of the parallel logic path. After all input and required instructions have been programmed in the last parallel branch, denote completion by the instruction **Branch End** and the keytop symbol shown in Figure 4-4.

Ladder Diagram Recommendations

Consider the following when constructing a ladder diagram rung:

- Any condition instruction can be repeated as many times as needed in a ladder diagram program or rung.

- Up to 12 condition instructions in series can be programmed in a rung.

- Only one output instruction can be programmed in a rung.

- Limit the number of parallel branch rungs to seven since this is the vertical limit of the Industrial Terminal.

- Program only one rung to control an output device (this simplifies troubleshooting and maximizes safety).

- When more than 12 condition instructions in series are required to energize an output device, use a storage bit and make two logic rungs.

- A logic rung is considered complete if it contains an output instruction.

- A program is generally considered complete if it contains the number of logic rungs required to operate a machine or process.

■ DEVELOPING THE PROGRAM

PRELIMINARY PROCEDURE

After constructing a ladder diagram, the next step is to establish an operating sequence for I/O devices. Evaluate the process to determine what the devices must do, the conditions under which they will work and the order in which they must operate.

After evaluating the operating sequence, it is helpful to sketch out the action of the different devices in their proper sequence with conditions for energizing each output device. This rough sketch can then used to develop the ladder diagram program.

Note: A relay ladder diagram cannot always be readily converted to PC ladder logic. Always consider the physical and mechanical characteristics of the I/O devices for each application before constructing a diagram.

PROGRAMMING PROCEDURE

The PC manufacturer's programming instructions should be followed step-by-step. The following is a typical example and refers to Figure 4-3:

- First, apply power to the industrial terminal.

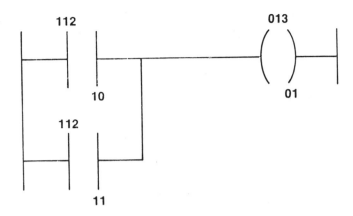

4-5. Ladder Logic Rung used by the Programmable Controller
Courtesy of Allen-Bradley

- Turn on the industrial terminal's power switch—this allows the CRT screen to display the instructions on the industrial terminal.

- Key MODE selection on the terminal keyboard. When the MODE selection message is displayed, press the bracket 1 key twice.

- Turn the mode select switch to the PROG position.

- Press SEARCH.

- Press CLEAR MEMORY; when memory is cleared, an END message will appear.

- Enter the relay logic diagram in the PC as ladder logic. (Figure 4-5 is an example of a rung which has two operating controllers in parallel, either of which can energize the output device.) Since this is a parallel circuit, enter BRANCH START first.

- Enter the first EXAMINE ON instruction. This EXAMINE ON instruction is identified as 112 10. This address refers to its location in the image table and also identifies the input module terminal which is wired to the operating controller. Enter the EXAMINE ON instruction for this rung by pressing the lower right key of the industrial terminal. Then give the symbol its numerical address 112 10. This insures that input module 112 10, which is electrically connected to address 112 10 in the input image table, will be scanned for the presence or absence of power in order to cycle the controlled device.

• Press BRANCH START again.

• Program the parallel EXAMINE ON instruction 112 11 in the same manner as 112 10 (see above).

• Since this completes the programming of the two controllers in parallel, press BRANCH END.

• The final step in this rung is to program the output device. This rung is complete when the output symbol on the lower right keyboard is pressed and numerically identified in memory. (The Allen-Bradley PC keyboard, Figure 4-3, does not require an END statement at the end of the rung. The processor automatically generates an END message when a new rung is started.)

In the foregoing manner, all the rungs and instructions for Figure 4-3 would be programmed in the PC. The PC manufacturer's individual procedures and instructions should serve as a technician's bible for successful programming.

■ CHECKOUT PROCEDURES

After programming the PC, certain precautions must be taken prior to an actual equipment start-up. These checkout procedures involve verifying proper voltage and power at all terminals. Make certain that all motors, valves and solenoids are disconnected until a checkout is completely finished. Also double check the inputs and outputs for errors in component identification and wiring. In addition, any program revisions should also be made at this time.

EDITING FUNCTIONS *(1)*

Function	Mode	Key Sequence	Description
Inserting a Condition	Program	[INSERT] [Instruction] [Address] Or	Position the cursor on the instruction that will precede the instruction to be inserted. Then press key sequence. *(2)*
Instruction		[INSERT] [⬅] [Instruction] [Address]	Position the cursor on the instruction that will follow the instruction to be inserted. Then press key sequence. *(1)*
Removing a Condition Instruction	Program	[REMOVE] [Instruction]	Position the cursor on instruction to be removed and press the key sequence.
Inserting a Rung	Program	[INSERT] RUNG	Position the cursor on any instruction in the preceding rung and press the key sequence. Enter the appropriate instructions to complete the rung.
Removing a Rung	Program	[REMOVE]	Position the cursor anywhere on the rung to be removed. **NOTE:** Only addresses corresponding to OUTPUT ENERGIZE, LATCH & UNLATCH instructions are cleared to zero when the rung is removed.

4-6. Example of Instructions for Editing Existing Programs
Courtesy of Allen-Bradley

(Continued)

Function	Mode	Key Sequence	Description
Change Data of a	Program	[INSERT] [Data]	Position the cursor on the word or block instruction whose data is to be changed. Press the key sequence.
		[CANCEL COMMAND]	To Terminate.
Change Data of a Word or Block Instruction ON–LINE	Run/Program	[SEARCH] [5] [1] [DATA]	Position the cursor on the word or block instruction whose data is to be changed.
		[INSERT]	Press [INSERT] to enter the new data into memory.
		[CANCEL COMMAND]	To terminate
Change the Address of a Word or Block Instruction	Program	[INSERT] [First Digit] [↓] [Addresses]	Position the cursor on a word or block instruction with data and press [INSERT]. Enter the first digit of the first data value of the instruction. Then use the [●] and [∅] key as needed to cursor up to the word address. Enter the appropriate digits of the word address.
		[CANCEL COMMAND]	To terminate.

(1) These functions can also be used during On-Line Programming.
(2) When bit address exceeds five digits, press the [EXPAND ADDR] key and enter a leading zero if necessary.

(4-6. Continued)

TESTING INPUTS & OUTPUTS

To check the I/O functions, turn the MODE select switch to TEST. Check all controllers for proper input module continuity and note if the proper rung displays appear. After all inputs are tested, test all the outputs. The manufacturer's manual should outline the recommended procedure for verifying proper output addresses and circuit continuity.

PROGRAM REVISIONS

Editing procedures allow changes to be made in an existing program. These procedures are usually outlined by the manufacturer. Figure 4-6 provides an example of one manufacturer's editing procedures. If a new rung needed to be inserted in the program, the example points out that the industrial terminal should be in the PROGRAM mode. The operator would then position the cursor on any instruction in the proceeding rung and strike the keys INSERT and RUNG. By following these steps, the rung is programmed in the same manner as the other rungs.

PC SYSTEM—FINAL CHECKOUT

Before final checkout it is imperative that controller connections and hardware have been tested and the ladder-logic program has been edited and validated. Checking machinery operation or processes should be accomplished with as little machine motion as possible. The best procedure is to test one machine output at a time and then make a final review of the ladder diagram. The manufacturer's manual recommends debugging procedures for complete program validation.

When the checks are complete, the final system checkout can begin. A typical final checkout is as follows:

1. Turn mode select to PROG.

2. Press SEARCH—this displays the first rung in memory.

3. Press CLEAR MEMORY—this clears memory and displays END statement.

4. Enter ladder diagram into memory.

5. Turn mode select switch to TEST.

6. Connect the output module to a single output device that starts machine operation.

7. Check the behavior of the foregoing output device. Energize the output and turn the processor mode select switch to RUN.

8. Disconnect the out output device.

9. Repeat steps 6—8 for each output device having machine motion.

10. Check to total application with all I/O devices connected.

After a satisfactory checkout of the entire system, it is recommended that the program be copied onto a cassette recorder as a security procedure.

MAINTENANCE & TROUBLESHOOTING

Cleanliness is a recommended preventive maintenance procedure. Periodic inspections help eliminate or minimize dust and dirt in the PC electrical system.

Troubleshooting should be a systematic, step-by-step operation. Manufacturers of PC equipment usually provide a troubleshooting flow chart which simplifies checks on circuits and operations by following a YES or NO structure. If a potential problem circuit or component is determined as YES, it is eliminated as a potential problem and the arrow leads to the next problem area. If a NO is determined for a component or circuit, the arrow points to suggestion blocks which help determine the cause and correction of the problem.

■ SUMMARY

This chapter on programmable controllers should reveal the relative simplicity and overall effectiveness of microprocessor-based control systems. PC's have many advantages. PC programming requires no specialized knowledge of computer assembly, languages, hardware or software. Also, in smaller and less complex applications, the PC is typically less costly. Also, the PC tolerates hostile environments better than a dedicated microcomputer can. The key to successful PC operation of hvac/r equipment is systematic and organized programming, editing and maintenance. Consistent performance enables any system to maximize equipment automation.

DDC PROGRAMMING
METHODS & CONFIGURATIONS

As applications become larger and more complex, the dedicated microcomputer system for hvac/r (hereafter referred to as the DDC system) becomes cost-competitive with the Programmable Controller (PC). For example, the relay logic of the PC (which is generally programmed with open and closed contact modes) becomes burdensome in larger applications. Complications are increased if the application requires timing or time-based decision making. Also, the PC utilizes individual circuitry while DDC controlled devices and system controllers can communicate over a single pair of twisted wires. Unlike the PC, the DDC system incorporates all control logic in the software. And since the DDC system is software driven, control set points do not require periodic recalibration. If control system values must be modified, PC's require a more cumbersome and time-consuming operation because the rungs involved must be re-programmed. In contrast, the DDC's menu-driven software in the English language allows setting changes and modifications to be made by simply positioning the cursor and striking a few keys. In these respects, the DDC is more 'user friendly' than the PC.

However, the PC and DDC systems both have definite individual application advantages. System choices are made by analyzing factors such as application size, cost, control complexity, the control capabilities of operating personnel and future plant or system requirements.

■ DDC SYSTEMS: CENTRAL & NETWORK CONFIGURATIONS

The DDC system can provide automation capability ranging from process control equipment to complete environmental applications for commercial, institutional or high-rise buildings. In its initial evolutionary phases, all hvac/r systems and sub-systems in a major application were under the jurisdiction of a **central** computer (Figure 5-1a). This computer provided total centralized surveillance and control. However, any failure or malfunction in the central computer affected and jeopardized the operation of all sub-systems and sometimes caused a shutdown of the entire hvac/r system. Obviously, a more stable system was necessary.

An improved arrangement is the **distributed control network** (Figure 5-1b). In this network, individual microcomputers perform a dedicated task in a zone or specific sub-system. In effect, each microcomputer is only distributed a portion of the total automation task. All microcomputers and the host (or supervisory) computer are linked together in a coordinated communication network. This configuration allows the host computer to perform major programs involving energy management, fire and safety, system security, centralized surveillance and retrieval and documentation of vital system information while each remote microcomputer performs its specific task as an independent **stand-alone control unit** (SCU). Consequently, any failure in the host computer or other sub-systems in the network does not affect the other computers.

Network capabilities can also be extended beyond the area of a major hvac/r system. Through telephone communication links, major systems can be networked to a centralized host computer. A **modem** (MODulation/DEModulation) is a chip or device capable of changing digital information to and from analog form. This enables computers and terminals to com-

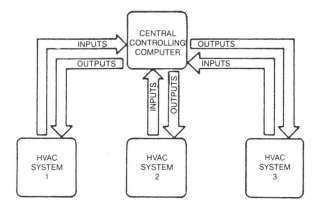

ALL SYSTEMS CONTROLLED BY A CENTRAL CONTROL COMPUTER

A. **Centralized control system.**

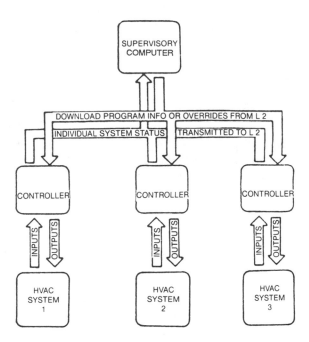

B. **Distributed control network.**

5-1. Centralized and Distributed Control Systems

municate information via telephone circuits. As a result, modems can link microcomputers and individuals in remote geographical locations.

DDC SOFTWARE

Software programs contain the step-by-step operation instructions for the microcomputer. Professionals who design and develop these instructions are called programmers. Their job is to translate the steps needed in operating an hvac/r system into language the microcomputer can understand.

This translation involves special language developed for computer technology. Language can be relatively simple or—if scientific technology is involved—very complex. The most common computer languages include FORTRAN (FORmula TRANsplation), COBOL (COmmon Business Oriented Language), BASIC (Beginner's All-purpose Symbolic Instruction Code) and APL (A Programming Language, a language for mathematical concepts).

THE FLOW CHART

A system analyst first gathers all pertinent data and programming information. The next step in software design is for the programmer to analyze the hvac/r control system. Questions to ask include: what are the operating objectives, and what functions must the controls and controlled devices perform to achieve the system objectives? After this system analysis, the programmer separates the control requirements into components. This is accomplished by a graphic step-by-step instruction format called the **flow chart**.

The flow chart is, in effect, a directional map for an operation which has a beginning and an end. All possible choices or steps and their consequences are shown in logical order. The microcomputer takes each step and processes it into a string of binary 1's and 0's so the microprocessor chip will initiate the required function.

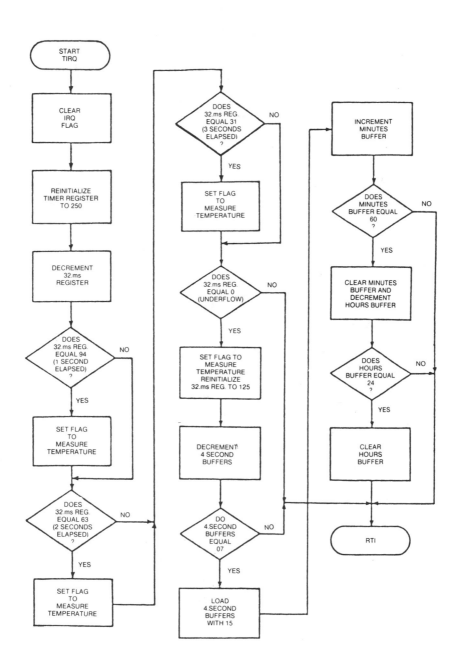

5-2. Example of a Programmer's Flow Chart Diagram

Figure 5-2 shows a flow chart involving the operation of a time clock. In a flow chart, boxes and various configurations of block symbols are used for particular steps. Directional arrows indicate the sequential flow of the functional operations in each step. Also note that the first symbol, an elongated oval, indicates the beginning. The elongated oval is the standard flow chart symbol for both the START and STOP functions of the program.

After START, the next three steps are symbolized by rectangles. Rectangles always represent an instruction. They give the microcomputer a specific instruction as a necessary step in the program's execution.

The next step, a DECISION, is symbolized by a diamond. A diamond always indicates a question and must be answered YES or NO. If the answer is YES, the microcomputer follows the YES arrow to the next box. If the answer is NO, the computer follows the NO arrow to its box.

The foregoing symbols are the most basic, but other symbols are used for various functions such as subroutines and connection points.

SUBROUTINE

A **subroutine** is a standard procedure which is applicable to many programs. Subroutines can be used repeatedly—for example, derivative and integral correction of proportional control sensing. Since the arithmetic operation is common to all proportional controllers, it can be branched into a small, independent program and assigned a specific location for future use after performing its assigned function. The symbol for a subroutine is a box with parallel lines on the top and bottom, and diagonal lines closing both sides. Another flow chart symbol, a small circle, identifies a connection point in the program.

THE PROGRAM

After a flow chart has been devised and checked for accuracy, the programmer uses it to code the instructions for the program. These step-by-step instructions must be in proper operating sequence. In addition, the chart must allocate memory locations for all data in the program. Each address must have an identification code or number. With the proper instruction, each pertinent data item can be selected (addressed) in much the same manner as dialing a telephone to communicate with a specific individual.

MENU

Because DDC systems are generally pre-programmed by the control manufacturer or hvac/r equipment manufacturer, the user rarely needs to program his DDC system. However, in the course of DDC systems operation, programs may require periodic modifications (i.e., changing set points, changes in timing, sequence alteration). As a result a method for user communication and input must be an integral part of the program.

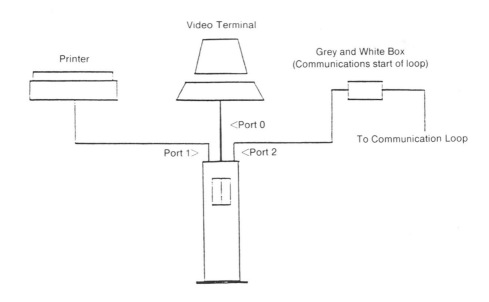

5-3. Basic Computer Control Terminal and Components
Courtesy of FES, Inc.

The user or plant operator must be able to communicate with the microcomputer in order to exercise necessary surveillance and supervision. This user instruction procedure is called the **menu**.

A menu displays to the operator all the available options on the CRT video terminal (Figure 5-3). The video terminal and keyboard shown allow the operator to complete system control from a single location. For example, if the system is a refrigeration plant, the software provides for the control and monitoring of all compressors and defrost zones. It can automatically sequence compressors, fans and pumps for maximum efficiency and economy. In addition it can provide a continuous update of plant machinery and also document system performance with hard-copy log reports.

Operation of the DDC refrigeration system in the previous example would not require any special computer knowledge. Menu-driven system commands involve only a few easy-to-learn keystrokes of the terminal keyboard. A well-designed program even provides checks against accidental or erroneous inputs or system failures.

KEYBOARD CONFIGURATION

Figure 5-4 illustrates an operator's keyboard for DDC operation of a refrigeration plant. The keyboard is designed for simplicity, and previous typing experience is not necessary. All **command keys** are located in the lower right corner, which is labeled the keypad area.

Next to the command keys are the **cursor keys**. The cursor indicates the location on the screen where data characters can be entered or corrected. In most terminals the cursor is as a flashing line just under the space where the selected keyboard character would appear when entered. Consequently the cursor location determines where printing or changes will start. By means of the cursor keyboard, the cursor can be positioned anywhere the operator dictates. The arrows on the cursor keyboard indicate the directional movements available to the key-

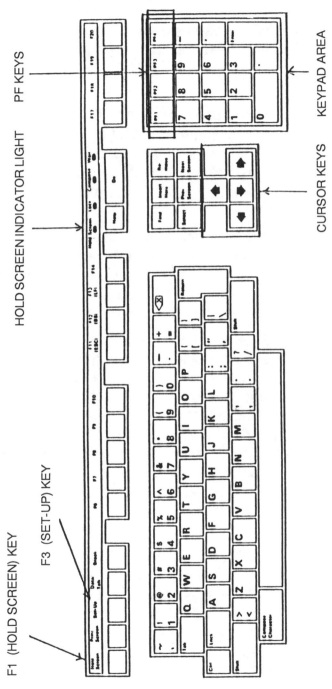

5-4. Operator's Keyboard for a Refrigeration Plant
Courtesy of FES, Inc.

board operator. With these arrow keys the operator can position the cursor anywhere on the display and thereby print anywhere on the display.

Figure 5-5 is an example of the screen display for evaporators in a menu-driven program. Operators with a substantial knowledge of refrigeration and of the requirements of their own plants should have no difficulty identifying components. In addition, the instructions on operations and setpoint options present no difficulty if followed literally and performed consistently.

In the Figure 5-4 keyboard only, the command keys and the cursor keys are on; the rest of the keyboard is turned off to prevent accidental strikes which could cause program errors or operational problems. The PF2 keys of the top row of the keypad area have their own command functions.

PF1 is the MASTER MENU key. When this key is struck, the system displays the master menu (Figure 5-6). The master menu enables the operator to select the specific screen that illustrates a component or the operating data needed for analysis. To make the selection, the operator types the number of the screen needed. For example, if the Suction Control Parameter Entry display is required, the operator types the #4 key in the keypad area. The operation is completed by striking the ENTER key (the elongated key on the extreme lower right of the keypad). The display shown in Figure 5-7 is then presented.

The entry SUCTION PRESSURE SET POINT indicates the PSIG pressure (the desirable suction pressure). SUCTION PRESSURE START POINT indicates the pressure which initiates a delay timer. If suction pressure should drop, the timer is cancelled; if pressure rises above set point, the compressor is started after the timer cycle is completed. This prevents a nuisance start-up due to a momentary pressure fluctuation.

PRESSURE STOP POINT is the pressure at or below which the compressor is stopped. A timer starts when this pres-

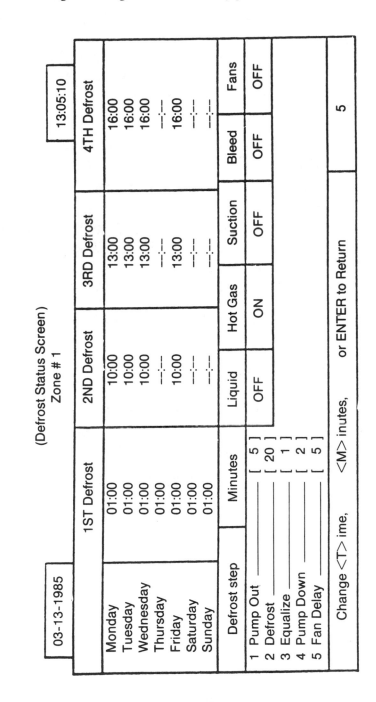

5-5. Evaporator Screen Display
Courtesy David A. Murphy, Frank Electric Corporation

sure is initially detected; if pressure should suddenly increase, the timer is cancelled. If pressure drops below the setting, the compressor starts after the timer completes its cycle. As a result, the timer prevents a nuisance shutdown due to a momentary low-pressure 'spike'.

The next two lines (T1 and T2) indicate the timing (in seconds) for START DELAY and STOP DELAY.

The T3 line, MINIMUM CAPACITY STOP DELAY, marks the time (in seconds) during which the system will keep a machine running at minimum capacity before stopping it. The line MAXIMUM SUCTION PRESSURE indicates the maximum that any machine's suction pressure may differ from the suction level average before an exception message is logged. Similar-

```
                         MASTER MENU

SYSTEM STATUS DISPLAY                               (1)

COMPRESSOR STATUS DISPLAY                           (2)

DISCHARGE CONTROL PARAMETER ENTRY                   (3)

SUCTION CONTROL PARAMETER ENTRY                     (4)

COMPRESSOR CONTROL PARAMETER ENTRY                  (5)

SUCTION PREALARM SETPOINT ENTRY                     (6)

COMPRESSOR PREALARM SETPOINT ENTRY                  (7)

DISCHARGE PREALARM SETPOINT ENTRY                   (8)

DISCHARGE CONTROL STATUS                            (9)

SET SYSTEM DATE AMD TIME                            (10)

ENTER LOG TIMES                                     (11)

DISPLAY EXCEPTION MESSAGES                          (12)

SAVE CONFIGURATION DATA                             (13)

DEFROST ZONE SETPOINT/STATUS                        (14)

DEFROST CONTROL STATUS                              (15)


ENTER SELECTION —
```

5-6. Master Menu Screen Display
Courtesy of FES, Inc.

ly, the MAXIMUM DISCHARGE PRESSURE DIFFERENCE is the maximum that any machine's discharge pressure can differ from the discharge average before an exception message is logged.

The usual procedure for making setting changes in the menu is to position the cursor on the setting location and then type in the new value. However, the procedure is subject to any pre-conditions the manufacturer may incorporate into the program. Consequently, each manufacturer's instructions should be noted and followed before attempting to program any set points or operating values.

When the operator has completed observations or changes on the data in a specific area, the MASTER MENU must again be displayed before another screen can be accessed. The return to the MASTER MENU is accomplished by striking the MENU key (PF1). Then the SYSTEM STATUS screen (Fig-

```
                    SUCTION PARAMETER ENTRY


    SUCTION LEVEL

       SUCTION PRESSURE SET POINT      (PSI)      2.0

       SUCTION PRESSURE, START         (PSI)      5.0

       SUCTION PRESSURE, STOP          (PSI)      1.0

       T1, COMPRESSOR START DELAY      (SEC)       30

       T2, COMPRESSOR STOP DELAY       (SEC)       30

       T3, MIN. CAPACITY STOP DELAY    (SEC)      300

       MAX SUCTION PRESSURE DIFF.                  6.0

       MAX DISCHARGE PRESSURE DIFF.               15.0
```

5-7. Suction Pressure Setting Screen Display
Courtesy of FES, Inc.

ure 5-8), COMPRESSOR STATUS DISPLAY or any screen in the master menu can be accessed by typing in its number and striking the ENTER key.

KEYBOARD OPERATION

The keyboard commands of the operating system are designed to simplify operator control. The operator does not need to be familiar with a conventional typewriter. All of the commands are located on the keypad area in the lower right corner of the keyboard, along with the cursor keys (Figure 5-4). The rest of the keyboard is turned off so that if any keys other than the keypad or cursor keys are accidentally hit, they will not cause any errors. However there are two exceptions: the F1 (Hold screen) and F3 (Setup) keys.

```
                    SYSTEM STATUS

    SUCTION LEVEL NO.                              **1**

       CURRENT PRESS.              PSIG            12.8

       SETPOINT                    PSIG            2.0

       PREALRM

    DISCHARGE LEVEL NO.                            **1**

       CURRENT PRESS.              PSIG            148

       SETPOINT                    PSIG            150

       PRELARM

    DISCHARGE CONTROL STATUS:    NORMAL

    DEFROST CONTROL STATUS:      2 ZONES IN ALARM


    MACHINE I.D. NO.    1   2   3

       RUNNING  (Y/N)   N   Y   N

       PREALARM

       FAILED                   Y


    RESET PREALARM (1), RESET FIRE ALARM (2)
```

5-8. System Status Screen Display
Courtesy FES Inc.

If the F1 key is mistakenly hit, it "freezes" the video screen so that no new information can be sent to the screen and no new commands can be accepted. If this happens, a red light turns on under the "hold screen" label (Figure 5-4). To "unfreeze" the screen, hit the F1 key once, The "hold screen" light will go out and screen activity will return to normal.

If an operator accidentally hits the F3 (Setup) key, the terminal goes into its setup mode. The screen is completely blank except for the "SET-UP Directory" message at the bottom. This command constructs the keyboard for different applications, which is equipped with the correction settings before it is sent to a job site. DO NOT change any of the keyboard attributes, as this renders the program inoperative. To escape from the "Setup" mode, simply hit the F3 (Setup) key once. This returns the screen to normal operation.

```
                    COMPRESSOR #1

  STOP          EXTERNAL    READY        OIL HEATERS OFF
  4             % SLIDE VALVE
  0             AMPS MOTOR CURRENT
  132           °F DISCHARGE TEMPERATURE
  114           °F INLET OIL TEMPERATURE
  114           °F OIL SEPERATER TEMPERATURE
  49            #'s OIL PRESSURE
  8             #'s OIL FILTER PRESSURE DROP
  155           #'s DISCHARGE PRESSURE
  5             #'s SUCTION PRESSURE


  AUTO(1), HOLD(2), LOAD(3), UNLOAD(4), EXTERNAL(5),
STOP/RESET(6), LOCAL(7), REMOTE(8), AR-TIME
REMAINING?(9), CLEAR ALARM(11), RESET PREALARMS(12)
```

5-9. No. 1 Compressor Screen Display
Courtesy FES Inc.

PF2 & PF3 Keys

The PF2 and PF3 keys are used on certain screens to step to the "next" or "previous" screen. For example: if a plant contains five compressors numbered 1 through 5, when the Compressor Status screen is first entered, the status of compressor #1 would be displayed (Figure 5-9). Pressing the PF2 key would display the following screen, compressor #2. If PF2 is typed again, compressor #3 would be displayed, and so on.

Alternatively, the PF3 key steps the screen back to the previous screen. If the screen shows compressor #3 and PF3 is typed, the screen steps to compressor #2. These keys also possess the "wrap-around" feature. For example, when stepping through the compressor screens in a plant having 5 compressors, the "next" screen after compressor #5 is compressor #1, and the "previous" screen before compressor #1 is compressor #5.

PF4 Key

The PF4 key is the PRINT key. Press the PF4 key at any time to obtain any one of three reports. This particular system presently has three different hard-copy log reports: the System Status Report, the Control Parameter Setpoints Report and the Prealarm Setpoint Report. The type of report provided depends on which screen the user is working. The System Status Report is printed when Status screens are employed, the Control Parameter Setpoints Report is given when using Control screens and the Prealarm Setpoint Report is printed when on Prealarm screens.

However, any one of these reports can be obtained by using the Master Menu. To do this, select the desired screen, press PF4 and press Enter. For example, if an operator wanted a hard-copy System Status Report while on the Master Menu, he would hit these keys: 1, 2 or 9 (Status screens), PF4 and Enter. A hard copy of the System Status Report would then be printed on the printer.

Cursor Keys

The Cursor keys are used by the operating system to position the current data entry point on the screen. This position is marked by using reverse video on the current entry point location. The cursor moves in the direction dictated by the arrows, but the program does not allow the cursor to move into an illegal area. The cursor also possesses the "wrap-around" feature. If the cursor is on the last entry point location on the screen when the down arrow is pressed, the cursor moves back to the top.

ENTER Key

The ENTER key is the elongated key on the lower right corner of the keypad. The ENTER key sends the command or data displayed on the screen to the computer. When making a menu selection, entering a parameter setpoint or issuing a command, press ENTER when the information is ready to be processed by the computer.

■ SUMMARY

With its centralized arithmetic-logic unit, microcomputer hardware has virtually infinite capability for precision control and operation of systems, from the smallest refrigeration plant to the largest factory, institution or high-rise building envisioned for the future. Moreover, all of these applications can be networked into multiples of hundreds and still be centrally controlled while maintaining the capability of stand-alone independent operation. The computer can readily cope with a variety of tasks, from routine daily system automation to more complex tasks such as initiation of permanent or temporary modifications necessary for energy efficiency.

At first glance the technology and responsibility of these systems may seem overwhelming. However, remember that the bulk of the responsibility of programming and operating systems really belongs to the programmer. The skilled programmer must develop reliable and tested programs for system operations. As for the technology, program instructions are interpreted into a "user friendly" menu format so that the average hvac/r technician should have little difficulty in using it to make required modifications.

DDC PROGRAMMING APPLICATIONS

The application of DDC hardware and operational software to conventional hvac/r systems requires a control technology transition that technicians must bridge. This section outlines and explains several typical system applications. Knowledge of these systems and their DDC operations will provide a better understanding of automation technology.

■ REFRIGERATED WAREHOUSE APPLICATION

The application details in this example represent a prominent manufacturer's method of DDC operation of evaporator systems in refrigerated warehouses.

Figure 6-1 represents an evaporator control configuration consisting of three temperature sensors and four solenoid valves. Temperature sensor A monitors air inlet temperature which in turn determines when the evaporator should stop or start. When air inlet temperature exceeds a pre-determined set point plus a dead band, the evaporator fan starts and circulates air in a stratification check.

After approximately five minutes, if the air is above set point plus dead band, sensor A energizes liquid line solenoid valve #1. Liquid in the coil initiates the refrigeration cycle, which

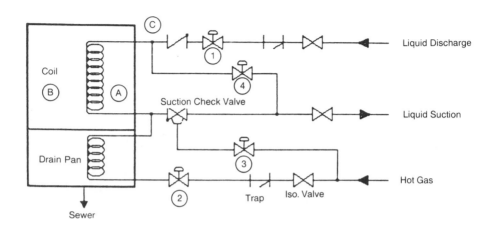

6-1. Refrigerated Warehouse Evaporator Control Arrangement
Courtesy of Pacific Micro-Control Corporation

continues until the air inlet temperature reaches the pre-determined low setting.

Temperature sensor B monitors air outlet temperature from the coil. As frost builds up on the coil, the differential between sensors A and B narrows and initiates the hot gas defrost cycle. It does this by turning off the fan, closing liquid solenoid valve #1, and opening hot gas solenoid valve #2, hot gas pilot solenoid #3 and crossover valve #4.

The hot gas defrost cycle terminates when temperature sensor C senses a temperature increase indicative of a completed defrost cycle. The suction hot gas pilot solenoid valve #3 opens during defrost to keep the suction check valve closed, which allows hot gas to pass directly through the coil. The hot gas then exits through hot gas relief regulator solenoid valve #4, which can be opened by an 85 psi pressure differential upstream of the liquid suction line. This solenoid valve can also be held open by being energized electrically.

Evaporator control of this application (Figure 6-1) could also be accomplished by using an electromechanical control system or solid state signal centers. However, doing so would not allow centralized energy management, system monitoring and surveillance, instant communication with many applications, and documentation of performance reports. These features could only be done manually, which would entail countless hours of manpower. Utilizing a microcomputer can overcome this limitation.

MICROCOMPUTER–BASED SYSTEMS

The dedicated microcomputer can automatically accomplish all the foregoing objectives for many evaporator applications when "daisy chained" into a network. A **daisy chain** network allows each unit to modify a signal before it is passed on to the next device. In addition, networks facilitate localization and diagnosis of maintenance and troubleshooting problems.

In a typical refrigerated warehouse application a stand alone control unit (SCU) can operate a number of evaporator systems through local multiplexers. The multiplexer takes in microcomputer temperature signals and emits electrical signals to open and close valves or start fans. Multiplexers can be connected in series through small–diameter, twisted, two-conductor cables with other multiplexers and to a host computer in a central control center.

Besides having a host computer, a monitoring control center can have a CRT display screen, a printer (for reports) and a modem. These peripheral devices allow the computer to communicate with other off-site computers. As a result it is possible to monitor conditions and change set points from home or a remote office. Also, the modem gives a host computer the capability of central control over refrigerated warehouses in other geographical areas.

Note: Multiplexer is an accepted industry reference for an I/O device which can have computer-working components and stand-alone capability.

The three temperature sensors in Figure 6-1 are connected by wiring to the input terminals of the multiplexer. The multiplexer in this application has the capability for 48 digital or 32 analog I/O points (or a combination of both). As a result other evaporator systems can be connected. The four evaporator system solenoids are connected to the output terminals of the multiplexer's I/O section.

After the software logic program has been entered into the central microcomputer, the system is ready for user operation. The multiplexer can function without intervention from the central microcomputer. This distributed control mode eliminates global problems associated with completely centralized computer systems.

In basic system operation of Figure 6-1, the software performs three main functions:

- Air inlet sensor A stops and starts the evaporator (based on the selected set point).

- Temperature drop across the evaporator initiates the defrost cycle (determined by differential temperature between air inlet sensor A and air outlet sensor B).

- Liquid/gas sensor C terminates defrost.

In addition, the software performs the following steps:

- Reads current analog inputs (temperature sensors) and digital inputs and outputs (switch positions) directly from all multiplexer units.

- Displays temperature readings on CRT screen.

- Stores current inputs and outputs in the computer memory.
- Applies control logic for most effective and energy efficient operation.
- Prints a periodic log.
- Monitors the refrigerant sensors and warns the operator when limits are approached. When limits are exceeded, it closes the appropriate valves to stop refrigerant flow.

These steps are repeated continuously, with each input being read at least once per minute. As the temperatures are scanned and displayed, they are compared to the program's pre-determined set points as selected by the user.

To comply with the fourth step (energy maximization), implement control logic. This method can control four different set point temperatures at different times of day, which could concentrate energy usage in off-peak periods when electrical rates are lower. Power surges (which may cause utility demand charges) can also be monitored and avoided. In addition, a program can limit the number of evaporators in a defrost cycle at one time (e.g., 30 %), and it can prevent an evaporator from defrosting more than once in a predetermined time period.

Upon completion of a cooling cycle, a fan loaded pumpdown (to boil off residual liquid) may occur before the fans are stopped. Before starting the evaporator, operate fans to reduce air stratification and assure representative room air temperature to the air inlet sensor. Permitting only one fan to start at a time minimizes power surges.

OPERATOR FUNCTIONS

The DDC system operator interfaces with the Evaporator Control System at the monitor and control center. The center consists of a host computer, a display screen, a keyboard, and often a modem and a printer. DDC manufacturers go to great lengths to provide user instruction manuals that are complete, concise and presented in a "user friendly" format. By following these instruction manuals the operator can request screen dis-

LIBRARY FUNCTIONS & ROUTINES

UTILITY FUNCTIONS:

PROCESS ENTRY	DISPLAY PANEL	DATA MANIPULATION	ALARM I.D.
PROG EVENT EVERY EXIT RESTART	DISPLAY ADJUST OVRIDE TIME DAY DATE YEAR ALMSCAN OVSCAN	STORE SET ACT BOUT AOUT FILE	ALARM DEVALARM

PROCESS ENTRY—A routine that allows entry to a process through the means of a PROGrammed time, an associated EVENT or on a start-up; RESTART. It also allows entry to a process on EVERY scan of the processor. Finally, every process will require an EXIT designation allowing the processor to scan the next process within the control program.

DISPLAY PANEL—These routines allow the control and display board the ability to DISPLAY and ADJUST data within the program as well as command overrides, OVRIDE, to output points. Other items that can be displayed are the current DAY, DATE, TIME and YEAR if desired. Whenever an alarm condition exists, an alarm scan, ALMSCAN, can be performed to determine which alarms are active. Any override commands that are active can also be determined by performing an override scan, OVSCAN.

DATA MANIPULATION—A set of routines that allow adjustment of binary, BOUT, and analog, AOUT, output commands as determined by the loop control routines. Data points can be SET to a specific binary status dependent upon what is desired. Data values are capable of being STOREd for use within the associated process or other related processes. Analog variables can be FILEd into memory and displayed at a later date, serving as a history function or Trend Log. Control Points, defined in the control program, can be ACTivated or deACTivated automatically within the control process.

ALARM I.D.—Allows ALARMs to be issued on the basis of unreliable input data or field hardware problems. If an analog variable deviates outside specified limits an alarm, DEVALARM, will be activated. All of these alarms can be displayed on the control and display board.

6-2. Programmed Software Routine
Courtesy of Johnson Controls, Inc.

MATH ROUTINES:

BASIC ARITHMETIC	BINARY LOGIC	RELATIONAL LOGIC	FIXED FORMULA
CALC SQRT POWER ABSOLUTE SIGN	OR SQR AND MOR TERM NEGATE	COMPARE ANDR ORR XORR	SELECT SPAN AVG TOTAL RATE RAMP ENTHALPY WETBLB WS RH DEWPT FILTER

BASIC ARITHMETIC—The CALCulation routine allows for addition, subtraction, multiplication and division of input variables. A square root (SQRT) or POWER of an inputed value can be performed as well as an ABSOLUTE value or a SIGN determination if necessary.

BINARY LOGIC—A collection of routines OR, XOR, AND that compare the status of binary variables, and on the basis of this status makes a decision allowing some function or event to take place. More than one OR routine can be used in succession prompting the use of a multiple term OR statement, MOR. A TERM routine is used in conjunction with the MOR statement to signify the TERMination of the MOR structure. The NEGATE function will invert the status of a binary variable.

RELATIONAL LOGIC—Analog input variables can be COMPAREd with one another. The ANDR, ORR, XORR functions, similar to the binary logic routines; AND, OR, XOR perform the compare statement and check the result of the comparison with the status of a binary variable. On the basis of this comparison and check, some function or event will take place.

FIXED FORMULA—These routines can be used in computations commonly required for HVAC control and Energy Management. They include psychrometric calculations of ENTHALPY, wetbulb (WETBLB), humidity ratio (WS), percent relative humidity (RH), as well as dewpoint (DEWPT) computation. Input variables can be computed and the highest or lowest value SELECTed and used within the control process. Input variables can also be SPANned over predefined limits to reset an input variable or setpoint used by the controller. An average (AVG) or TOTAL can easily be computed by the DSC-8500 controller to perform a gradual start-up of mechanical equipment. The FILTER routine can be used to smooth out an input signal that is in constant fluctuation. An example would be static pressure on a variable air volume system.

6-3. Programmed Math Routines
Courtesy of Johnson Controls, Inc.

CONTROL ROUTINES:

COMPENSATION	LOOP CONTROL	ENERGY CONSERVATION	TIMED PROGRAM
LEDLAG HYS	PROP INCR MODE3	DUTY ESTTC LOAD SUPPLY	DELAY HOLIDAY INTERVAL TIMTOT TIMDATA TIMRESET

COMPENSATION—The LEDLAG routine is a compensation calculation performed by the controller where special signal conditioning is required for improved control performance. A HYSteresis routine can be used to compensate for hysteresis in the controlled device.

LOOP CONTROL—The PROP routine provides proportional, proportional plus integral, or proportional plus integral plus derivative control (or in any desired combination). INCRemental control is another form of PID control that measures the velocity and acceleration of changes in the controlled variable and also utilizes a deadband which is considered zero error.

ENERGY CONSERVATION—DUTY cycling can be performed using the microcomputer as a stand-alone or when used in conjunction with a BAS system. The estimated process time constant, ESTTC, can be used to estimate time constants for Optimal Run Time applications. For optimal state appliations, the SUPPLY AND LOAD routines can be used to determine optimum system start time for building warm-up or cool-down.

TIMED PROGRAM—These routines can provide a timed DELAY or INTERVAL within the respective processes. Provisions for system operation/shutdown during HOLIDAYs can be made. Run time totalization, TIMTOT, can be performed any time total equipment runtime is desired. TIMDATA allows the current year, month & day, day, day-of-the-week, or time of day to be used within a specific control process. TIMRESET can be used in conjunction with TIMDATA in order to reset the value used at TIMDATA, as in daylight savings.

These routines, with their inherent flexibility within the microcomputer application program, allow the programmer the ability to control a wide array of HVAC systems that also include Energy Management and Diagnostic functions.

6-4. Programmed Control Routines
Courtesy of Johnson Controls, Inc.

plays which indicate current operating conditions and system parameters.

Set points and time delay parameters can be changed by a simple keyboard procedure as outlined in the user's manual. In addition, commands may be manually transmitted to the multiplexers. One of three modes of operation may be selected: the preprogrammed Demand Mode (designed for completely automatic operation and energy management), the Schedule Mode or the Manual Mode.

DEMAND MODE

To operate defrostable evaporators in the Demand Mode, the operator must specify the following parameters:

Room Air Temperature Control:

Room air temperature set point (RAT)

Deadband (DB)

Circulation time for stratification check (t_S)

Delay starting fans before opening liquid valve (t_F)

Fan loaded pumpdown period (t_P)

Defrost Control:

Minimum cooling period before defrost test (t_C)

Coil air temperature drop set point (CATD)

Delay after opening hot gas pilot before opening hot gas valve (t_A)

Delay after closing hot gas valve before closing hot gas pilot (bleed period) (t_B)

Hot gas exit temperature set point (HGET)

Maximum defrost cycle time (Max_D)

RAT, DB and CATD may be specified independently by time of day for up to four time periods. The remaining parameters are constants applicable to all time periods.

For room air temperature control, the deadband is applied above the set point, and temperature is controlled between RAT and (RAT+DB). If the temperature measured by an evaporator's inlet air sensor is above (RAT+DB), that evaporator's fan is activated to circulate the air in a stratification check. Circulation continues for tS minutes. At the end of this time, if room air temperature is still above (RAT+DB), cooling is initiated.

Cooling is terminated if the inlet air temperature falls below RAT while cooling is in progress. Normal termination of the cooling cycle involves closing the liquid solenoid valve while letting the fans run for tp minutes to boil off residual liquid (this is called a fan loaded pumpdown).

For each evaporator, if a cooling cycle has been active for at least tC minutes, the temperature drop across the evaporator is determined by subtracting air outlet temperature from air inlet temperature. This difference is adjusted using a heat balance based upon the measured coil liquid temperature. If the adjusted temperature drop is less than the evaporator air temperature drop set point CATD (indicating that cooling efficiency is below par), that coil becomes a candidate for defrost. If defrost demand is indicated continuously for 2 x tC minutes, the cooling cycle is terminated (with fan loaded pumpdown) and a defrost cycle is initiated.

During the defrost cycle, hot gas flows through the evaporator in the reverse direction, giving up some of its heat to melt the frost. As frost is dissipated, the temperature of gas leaving the evaporator rises. When this hot gas exit temperature becomes greater than its set point HGET, the defrost cycle is terminated. Alternatively, if defrosting continues for its maximum time limit MaxD before hot gas exit temperature reaches HGET, the defrost cycle is also terminated.

MANUAL MODE

System control of solenoid valves and fans is performed by sending discrete output signals to the control multiplexer units. Note that Manual Mode commands supersede Demand Mode signals. Because the multiplexer communicates using standard ASCII characters, it is possible to send commands and receive replies using a video display terminal (VDT) regardless of computer status.

When the computer is operative, commands may be directed through the computer from the keyboard to the multiplexer. The computer will receive and display each reply. When the computer is inoperative, a switch permits the VDT to be disconnected from the computer and reconnected to the multiplexer network. In this mode, commands typed on the keyboard will be transmitted directly to the multiplexer and each response will appear on the video screen.

SCHEDULE MODE

In addition to the Demand Mode and the Manual Mode, the Schedule Mode offers a method of operating the evaporator system in the storage and loading dock areas. Schedule Mode software logic can initiate three different actions up to 20 different times per day. These actions are: Begin Cooling, End Cooling and Start Defrost (Stop Defrost is controlled by the variables HGET and MaxD). The operator may establish up to ten different daily action schedules and select one for application to an individual evaporator by merely inserting a schedule-identifying number on the keyboard of the Monitoring and Control Center.

There are some advantages to the Schedule Mode. The Schedule Mode allows a specific action for each evaporator in the storage or loading dock areas to be performed for a fixed period at a specific time each day. This method of operation is especially useful when running a series of system performance tests or observing the performance of a specific evaporator during several cycles.

In addition, instead of manually performing several actions on individual components to start or stop a cycle, Schedule Mode software logic automatically carries out designated actions at designated times. The Schedule Mode also provides an easily implemented semi-automatic mode of operation as a back-up to the Demand Mode, which is sometimes temporarily inoperative due to repair or adjustment of temperature probes.

Schedule Mode Software Logic

The Schedule Mode requires no additional hardware, but its software logic does have certain programming requirements which should be noted.

The following control actions may be specified to occur according to a time schedule:

Schedule Mode Control Actions	
Begin Cooling	(BC)
End Cooling	(EC)
Begin Defrost	(BD)

There is no need for a command to schedule an end to hot gas defrost because it terminates automatically whenever the hot gas exit temperature reaches its set point HGET or when the defrost cycle has exceeded its maximum time limit MaxD.

The operator may prepare up to ten different scheduled actions at the video display terminal. For each action, the operator must define the time (in 24-hour clock format) and action codes (BC, EC or BD) until the schedule is fully described. For these ten actions, up to 20 action codes may be specified for each schedule.

For example, a schedule to automatically begin cooling at midnight, defrost at 6:00 a.m. and cool until 10:00 p.m. would read as follows:

0000	BC
0545	EC
0600	BD
0630	BC
2200	EC

If the schedule action is Begin Cooling (BC), a cooling cycle starts immediately without going through an air stratification check circulation period. If the scheduled action is End Cooling (EC), a normal fan loaded pumpdown period occurs.

Evaporators which are scheduled to Begin Defrost (BD) are first allowed to end cooling with fan loaded pumpdown if the pumpdown was not previously completed. If a coil is switched from Schedule Mode to Demand Mode or Manual Mode before beginning a pending defrost cycle, its scheduled defrost action is overridden by the new control mode. Once a defrost cycle starts, it is allowed to terminate normally (based upon HGET or MaxD) before the next scheduled action takes effect.

Note: on non-defrostable evaporators (such as usually found on the loading dock), Begin Defrost does not apply.

BASIC CONTROL SYSTEM SELF-DIAGNOSTIC CAPABILITY

A basic control system is composed of a few high-quality components which are used because of their dependable, trouble-free operating history. Because of the modular design of multiplexers and the computer, a few well-chosen spare modules can allow local maintenance to correct malfunctions.

The basic control system has been designed to allow the cause of any malfunction to be diagnosed in a simple, straightforward manner. For example, a malfunction in the following modules would be diagnosed as follows:

1. Power supply: if it is not possible to communicate with any multiplexer boards in a single unit, then the power supply is assumed to be faulty.

2. Multiplexer Board: if it is not possible to communicate with any points on a single board, but other boards in the same unit are responding, then the specific multiplexer board is assumed to be faulty.

3. Input/Output Modules: when an invalid analog module is addressed, an error code is returned. If this occurs three times in a row, then the module is assumed to be faulty.

Faulty digital modules do not return a similar code. One way to confirm proper digital module operation is to verify that the expected state is present. For example, if an ON state has just been output, then an ON state should be expected for the next input.

4. Temperature Sensors: If the signal is above or below reasonable limits, a faulty sensor is indicated.

5. Communication Link: If the software cannot communicate with any of the multiplexer units, then the serial network link is assumed to be faulty.

■ HVAC APPLICATIONS

Pneumatic and solid state control systems and centralized control centers have the logic capabilities to effectively operate individual hvac systems. However, for these to have the ability to attain the control refinements required for automation would involve a multiplicity of hard-wired components. Even so, they could never achieve the versatility of the most basic DDC system. In addition, the conventional systems would lack the ability to communicate and coordinate with other applications in centralized operations.

The overwhelming difference in capability between systems is due to software. Software enables DDC systems to have unlimited options for modification and communication, a requirement for total automation.

LIBRARY FUNCTIONS

Although software is the basis of effective DDC systems, the programming language must be easy enough for the operator to use and understand. Consequently, DDC manufacturers try to program the routines needed for hvac processes in an English language format which meets the needs for user comprehension and system operation. These processes or routines may be found in a **library**, a collection of standard programs and subroutines used in problem-solving.

Figures 6-2, 6-3 and 6-4 list established control manufacturer's DDC library functions. These functions involve all the routines programmed in ROM required to automate a hvac system and make it completely operational. The processes that can be activated and accomplished by these library functions include:

- Energy management and conservation
- Time programming
- Control and display board operation
- Process initiation and termination
- Alarm identification
- Arithmetic operation
- Binary and relation logic
- Formula computations
- Control signal compensation
- Closed loop control

As the following basic application example points out, each DDC routine has an associated parameter listing which determines how the routine will function within the selected process. In short, the parameters contain the specific values and instructions needed for effective system operation.

The library functions or routines are grouped into three categories: utility functions, math functions and control routines. The individual functions within each category are thoroughly explained in Figures 6-2, 6-3 and 6-4.

Field Points

Terminal connections for sensor inputs as well as transducer and controlled device outputs are made at well-defined locations in the I/O unit. These locations are called **field points**. Typical programmable field points for an hvac system would be as follows:

PROGRAMMABLE FIELD POINTS	
CONTROL POINT	**ANALOG INPUT**
Analog Interlock	Full Range
Binary Interlock	Limited Range
Binary Output w/Feedback	Analog Rate
Incremental	Analog Total
BINARY INPUT	**BINARY OUTPUT**
Contact Status	Maintained Output
Pulse Rate	Momentary Output
Count Totalization	Position Output

Data Points

Data points in the control program contain binary or analog information and serve as temporary or permanent storage locations. These defined data points for binary and analog variables are required when data is to be recorded and stored for future reference. Data points can be used as input or output values to a process or processes within a program. These points can be modified through trunk line communications by means of the control and display board adjustments of the microcomputer.

An example of data point listing and explanation would be displayed as follows:

DATA DEFINITION POINT			
		P1	P2
BD-1	CLG	D	R
AD-1	CLGSP	D	55.0

BD-1 defines the data point as binary data point #1. CLG stands for Cooling. P1 and P2 are the parameters to which the data point must conform. P1 contains the commands **Enable (E)** or **Disable (D)**, which determine whether or not information will be communicated. An E (go) command would cause a cooling event status to be reported to a host computer or other appropriate display. In the example, however, the command in P1 is D, a stop communication command. Consequently, parameter P2 displays **Reset (R)**, the designation for OFF. The message in the example is that system cooling is not required.

In the second data point, analog data point #1 (AD-1), **CLGSP** refers to cooling set point. In parameter P1, the D symbol signals that no cooling is required. However, parameter P2 displays 55.0, which is the set point (in degrees Fahrenheit) to be maintained when the cooling system is in operation.

SOFTWARE CONTROL LOOP

Figure 6-5 illustrates how system software can link defined hardware points for a controlled device (such as a damper or valve) into a means for a new position as demanded by the controller.

CONTROL AND DISPLAY BOARD

Figure 6-6 is a control manufacturer's **control and display board (CDB)**. This board is a typical stand alone microcomputer control and display device which allows the operator local access and program control of the DDC system. With the CDB, analog and binary values can be displayed and controller variables can be adjusted. In short, the user can establish and maintain control strategies, tune the system, change time schedules, override routines and make all necessary adjustments and corrections.

In addition, statements of system status can be displayed to inform maintenance personnel on specifics for localizing and solving troubleshooting problems. Complete information on system management and energy usage is immediately available from a keypad format. Moreover, the data can be documented with the addition of a printer.

The basic programmable functions for hvac system operation are shown in Figure 6-7. Keep in mind that these do not apply to every situation. Before programming a system, always consult the manufacturer's operating manual for detailed operating instructions and procedures.

Figure 6-8 illustrates a DDC control application for a basic central fan system. The **discharge air temperature sensor (DISTE)** is a resistance change device which sends an electronic signal to an **electronic-to-pressure (E/P) transducer**. The transducer converts the electronic signal to pneumatic pressure which actuates the **cooling valve (CLVLV)**.

Figure 6-8 also delineates general field points, data points, and CDB functions needed to monitor and adjust the system. The following table defines all the operating parameters associated with each field device. Figure 6-5 depicts how the **control point (CP**-1) receives instruction from the PCR to reposition the chilled water valve from A1-2 to B0-1.

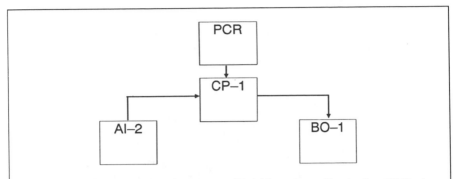

A control point, resident in the Field Interface Controller (FIC), is a software point capable of linking together defined hardware points. With respect to the cooling loop, the control point will link the position feedback point (AI-2) with the binary output point (BO-1).

The control point receives the required output command from the processor board (PCR). Based on this information, the control point will refer to the actual position of the valve (AI-2) and then calculate the number of output pulses (from 0 to +/-255) that are required to drive the valve from its present position (AI-2) to the desired position (BO-1). The parameters associated with the incremental control point (as it is called) allow for communication with a BAS system and/or between the FIC and PCR. It also describes the names of the points that are to be linked together as well as tuning coefficients necessary for a properly working control system. These tuning coefficients are established during the program design and can be modified as desired at the controller location.

The binary output point (BO-1) receives the required number of output pulses from the incremental control point and assigns a time, in milliseconds, to each pulse it receives. The result is a +/− 12V DC signal applied to the actuator for the duration of time determined by the binary output point. The number of pulses multiplied by the time base defined at the binary output point will determine the length of time that the +/− 12V DC signal will be applied to the actuator. The various times that can be assigned by the binary output points are listed below.

P3	Milliseconds/Count
0	20
1	60
2	120
3	240

6-5. Software Control Loop
Courtesy of Johnson Controls, Inc.

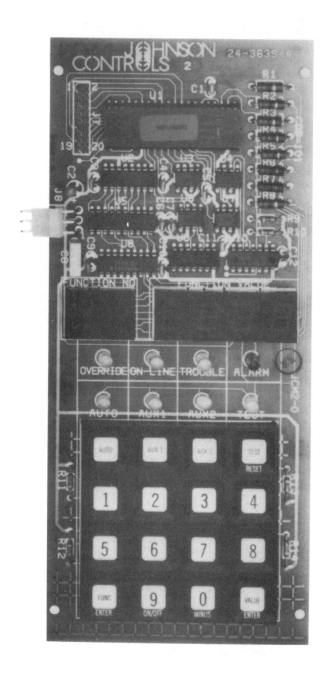

6-6. Microcomputer Control And Display Board
Courtesy of Johnson Controls, Inc.

CONTROL AND DISPLAY BOARD

The control and display board allows for display and manual adjustment of predefined functions within the control program and provides the local operator interface to the control system. When used in conjunction with the processor board, the control and display board can be used to display binary and analog values, adjust controller variables, display control system alarms, and provide a manual means of control if desired. The control and display board is programmable, allowing for 100 specific functions custom-fit to each particular system that the microcomputer is applied to. The basic programmable functions are listed below. (Detailed procedures for individual systems are listed in each manufacturer's operating manual.)

PROGRAMMABLE FUNCTIONS	
DISPLAY	Display a selected analog or binary variable
ADJUST	Display and allow adjustment of a selected analog or binary variable
OVRIDE	Provide manual override control of a selected output point at programmed authority
TIME	Display and allow adjustment of time-of-day
DAY	Dislay and allow adjustment of day-of-week
DATE	Display and allow adjustment of month and day
YEAR	Display and allow adjustment of year
OVSCAN	Display and allow reset of manual override functions

6-7. Programmable Functions of the Control & Display Board
Courtesy of Johnson Controls, Inc.

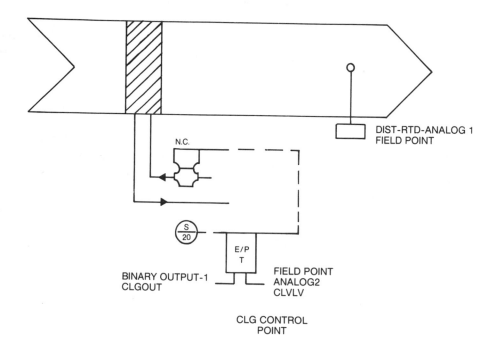

DIST-RTD-ANALOG 1
FIELD POINT

N.C.

BINARY OUTPUT-1
CLGOUT

E/P
T

FIELD POINT
ANALOG2
CLVLV

CLG CONTROL
POINT

DATA POINT DEFINITIONS:
BD-1	CLG	RESET
AD-1	CLGSP	55°

FIELD POINT DEFINITIONS:
AL-1	DISTE	FUL
AL-2	CLGVLV	FUL
BO-1	CLGOUT	POS
CP-1	CLGCP	INC

CONTROL AND DISPLAY BOARD FUNCTION CODE:
0	ADJUST	CLGSP	AL
1 & 2	DISPLAY	DISTE & CLGVLV	AL
3	OVRIDE	CLGCP	PRIORITY 2

6-8. Control Points in a Central Fan System

Courtesy of Johnson Controls, Inc.

COOLING DISCHARGE CONTROL PROGRAM

Field Point Definition

			P1	P2	P3	P4	P5	P6	P7	P8	P9
			Field Points								
AI-1	DISTE	FUL	E	1.0	E	V	T	-50.0	233.4		
AI-2	CLGVLV	FUL	D	1.0	E	N	O	-12.5	250.0		
BO-1	CLGOUT	POS	D	E	O						
CP-1	CLGSP	INC	D	E	A	CLGVLV	CLGOUT	-140	0	5	0

The defined field points for the cooling control loop describe pertinent information regarding the **Field Termination Board (FTB)** wiring location and the function of each defined point. Two analog input points, AI-1 and AI-2, are defined for this program.

Analog input#1 (AI-1) defines a temperature element (DISTE) located at terminal group #1 for the analog input section on the FTB-102. An associated designation, FUL, indicates that the entire designated sensed temperature range (-50.0° to 233.4° F.) will be utilized by the controller. A resistance value equal to particular sensed temperature is obtained by the controller. This value is then converted to a digital signal that can either be displayed as temperature on the CDB, reported to a **Building Automation System (BAS)**, used to monitor the microcomputer or used by a control process within the PCR-102.

Because the microcomputer is capable of becoming an integral part of a computerized building automation system, certain parameters must be properly defined so that proper communication occurs between the devices.

The parameters associated with AI-1 which allow this communication are P1 and P2. These parameters will determine if and when inter-device communication will take place. The parameter listing for P1 is an E (Enable), indicating that PCR-102 will report a new data value (temperature) to the BAS whenever the data value changes by more than the designated filter increment P2.

In this example the report occurs only if the discharge temperature deviates by +/-1.0 F since the last report was issued to the BAS. If no communication with a BAS was desired, a D (Disable) would be placed at P1.

Parameter #3 (P3) indicates whether or not communication will take place between the Field Interface Controller (FIC) and the PCR using the associated values of Disable or Enable. Most applications require the use of Enable since it is the PCR that contains the control program and requires field input provided by the FIC.

Resistance-to-temperature devices (RTD's) require the use of the 5.5V DC analog voltage supplied by the RPA-101. It is possible that a slight drift in this 5.5V DC supply may occur, and this affects the RTD's accuracy. Parameter #4 (P4) allows for voltage compensation in the event that drift takes place. The required designations for P4 can be V (Voltage correction required) or N (No voltage correction) and are necessary when potentiometers or voltage inputs are used.

Its required values are determined by P4's designation. If any RTD is used as an input, the proper designation for P5 will be T. If anything other than an RTD is used as an input, then the proper designation for P5 will be the letter O. As a result this statement can be made:

if P4 = V then P5 = T, and if P4 = N then P5 = O.

The proper designations of P4 and P5 insure that accuracy is maintained during the operation.

The final parameters for the analog input point are P6 and P7, calibration coefficients which determine the range of values over which the controller will operate. The values can be used by the CDB-101, the BAS or the PCR-102. The values for P6 and P7 are also used as compensation in software for variations in hardware such as RTD's, field cable length and FIC components. These values can be calculated through a mathematical formula or by using the label attached to the FIC in use and are entered when writing the application program.

Analog Input two (AI-2) takes on the same number of parameters and value options for each of the parameters. This particular point, CLGVLV, is defined in order to monitor the position of the cooling valve. This valve position, if read on the CDB-101, is displayed as a number from 0 to 100.0. These numbers represent the range of valve travel based on a scale of 0% open to 100% open.

The input to AI-2 would come from a potentiometer, part of an **electric-to-pneumatic transducer** (**EPT**-102). This results in P4 and P5 designations of N and O, respectively.

A binary output must be defined in order to drive the controlled device to the position required by the microcomputer (see Figure 6-5).

DDC SINGLE ZONE CENTRAL FAN SYSTEM

Figure 6-9 illustrates the basic configuration for a single state heating and cooling central fan system with economizer refinements. The microcomputer controller's analog inputs (Figure 6-10) can be used as adjustable set points when wired to a panel of potentiometers. Binary inputs are also provided.

The controller communicates with I/O devices in the system over a Level 1 bus (a pair of twisted wires). The controlled devices receive commands from the controller and continuously respond with reports of their actions. If a controlled device

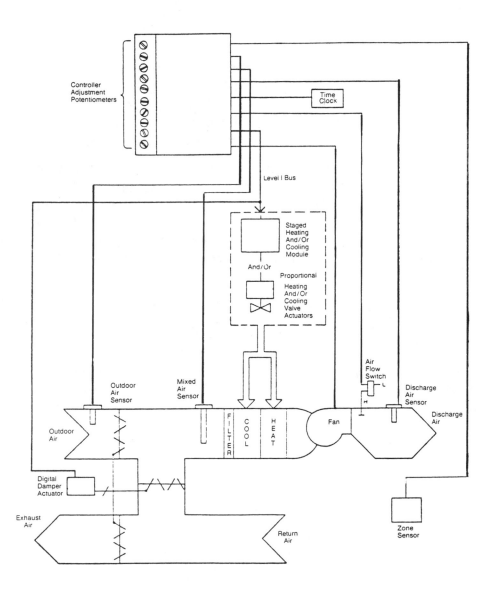

6-9. Stand Alone Control for a Central Fan System
Courtesy of Johnson Controls, Inc.

does not respond to the controller's command, an error message is displayed.

The system in Figure 6-9 is a complete independent operation for a central fan system. If it were part of a network of central fan systems, it would communicate with the host computer and other stand alone systems through the Level 2 pair of wires at terminals 29 and 30 in Figure 6-10.

Figure 6-9 also illustrates how a modular sub-base can activate relays to provide stages of heating and cooling. If proportional control response is required, a digital-to-proportional interface device makes the conversion for modulated performance of a motor actuator. For pneumatic actuator performance, a pneumatic interface device must be added.

The factory-installed software dedicates the microcomputer controller to perform the functions needed for the specific application. In the case of a single-zone central fan system, check Figure 6-11 for the heating mode operational sequences. Figure 6-12 covers the cooling mode, while Figure 6-13 involves the economizer cycle.

■ SUMMARY

The applications covered represent the hvac/r industry's approach to DDC systems. Although software is fundamental to control systems, the operator has considerable input regarding system performance. For example, the operator can monitor sensor readings and readjust set points and controlled device positions. In addition, the operator has centralized authority to localize service problems and to implement remedial responses. This perspective on how DDC hardware and software function in conventional applications is intended to help clarify the automation techniques and configurations discussed in the next section.

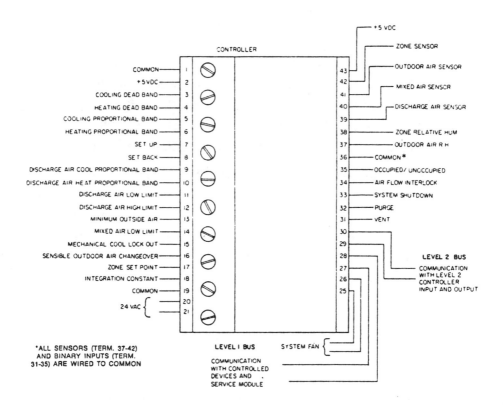

CONTROLLER

Left Terminals		Right Terminals

+5 VDC

COMMON — 1
+5 VDC — 2
COOLING DEAD BAND — 3
HEATING DEAD BAND — 4
COOLING PROPORTIONAL BAND — 5
HEATING PROPORTIONAL BAND — 6
SET UP — 7
SET BACK — 8
DISCHARGE AIR COOL PROPORTIONAL BAND — 9
DISCHARGE AIR HEAT PROPORTIONAL BAND — 10
DISCHARGE AIR LOW LIMIT — 11
DISCHARGE AIR HIGH LIMIT — 12
MINIMUM OUTSIDE AIR — 13
MIXED AIR LOW LIMIT — 14
MECHANICAL COOL LOCK OUT — 15
SENSIBLE OUTDOOR AIR CHANGEOVER — 16
ZONE SET POINT — 17
INTEGRATION CONSTANT — 18
COMMON — 19
24 VAC { 20 / 21

43 — ZONE SENSOR
42 — OUTDOOR AIR SENSOR
41 — MIXED AIR SENSOR
40 — DISCHARGE AIR SENSOR
39
38 — ZONE RELATIVE HUM
37 — OUTDOOR AIR R H
36 — COMMON *
35 — OCCUPIED/ UNOCCUPIED
34 — AIR FLOW INTERLOCK
33 — SYSTEM SHUTDOWN
32 — PURGE
31 — VENT
30
29
28
27
26
25

LEVEL 2 BUS

COMMUNICATION
WITH LEVEL 2
CONTROLLER
INPUT AND OUTPUT

*ALL SENSORS (TERM. 37-42)
AND BINARY INPUTS (TERM.
31-35) ARE WIRED TO COMMON

LEVEL I BUS

COMMUNICATION
WITH CONTROLLED
DEVICES AND
SERVICE MODULE

SYSTEM FAN {

6-10. Fan System Control Panel
Courtesy of Johnson Controls, Inc

SEQUENCE OF OPERATION

Call for Heating

1. As zone temperature drops from the set point, no mechanical heating will occur as long as the zone temperature remains within the selected heating dead band.

2. As the zone temperature decreases to the point where it enters the heating proportional band, the controller calculates the discharge air temperature reset ramp using four coordinates:

- "A" the discharge air temperature at the point where the zone temperature enters the heating proportional band.

- "B" the highest zone temperature point within the heating proportional band.

- "C" the discharge air high limit setting.

- "D" the lowest zone temperature point within the heating proportional band.

3. As the zone temperature continues to drop through the heating proportional band, the discharge air temperature set point will be reset upwards along the heating ramp calculated by the controller.

4. As the heating demand decreases, the discharge air temperature set point will decrease, following the calculated heating ramp.

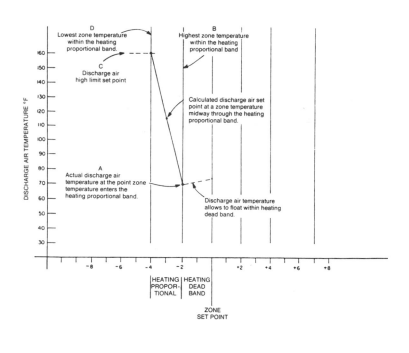

6-11. Operating Sequence for a Central Fan System - Heating

Courtesy of Johnson Controls, Inc

Call for Mechanical Cooling

1. As zone temperature increases to the point where it enters the cooling proportional band, the microcomputer or controller calculates the discharge air temperature reset ramp using four coordinates:

- "A" the discharge air temperature at the point where the zone temperature enters the cooling proportional band.

- "B" the lowest zone temperature point within the cooling proportional band.

- "C" the discharge air low limit setting.

- "D" the highest zone temperature point within the heating proportional band.

2. As the zone temperature continues to drop through the heating proportional band, the discharge air temperature set point will be reset downwards along the cooling ramp calculated by the controller.

3. As the cooling demand decreases, the discharge air temperature set point will increase, following the calculated cooling ramp.

Mechanical Cooling Lockout

When the outdoor air temperature is below the selected lockout set point, mechanical cooling is not allowed to operate. This reduces needless compressor operation or possible damage to the compressor.

6-12. Cooling Sequence for Central Fan System

Courtesy of Johnson Controls, Inc

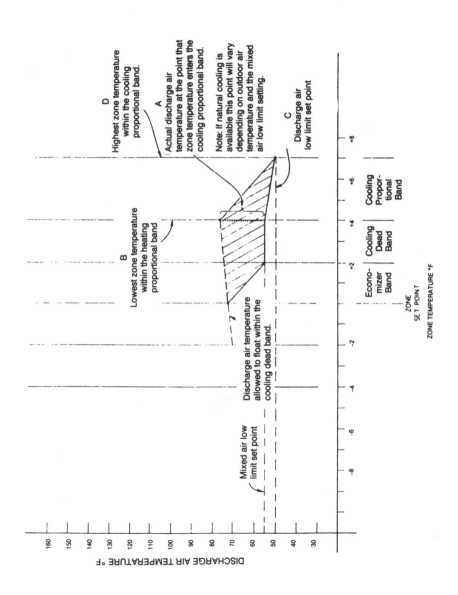

(6-12. Continued)

Call for Economizer

1. As zone temperature rises through the first 2°F above the set point, the economizer function will drive the outdoor air damper from its minimum position to its full open position. During the first two degrees above zone set point, the system will attempt to use outdoor air to satisfy the cooling requirements of the zone if:

    (a) natural cooling is available

    (b) the mixed air temperature is above the mixed air low limit set point.

2. As the zone temperature increases beyond the two degree economizer band, the system is allowed to "float" within the cooling dead band.

Note: the cooling dead band starts at the end of the economizer proportional band, even if natural cooling is not available.

Mixed Air Low Limit

This function proportionally overrides the economizer actuator to prevent the mixed air temperature from dropping below the mixed air low limit set point.

The mixed air low limit proportional band setting is automatically increased (4,8 or 16°F) as the outdoor air temperature drops. This function keeps the outdoor air damper motor actuator from "hunting" when there is a call for cooling and the outdoor air temperature is low.

6-13. Economizer Operating Sequences
Courtesy of Johnson Controls, Inc

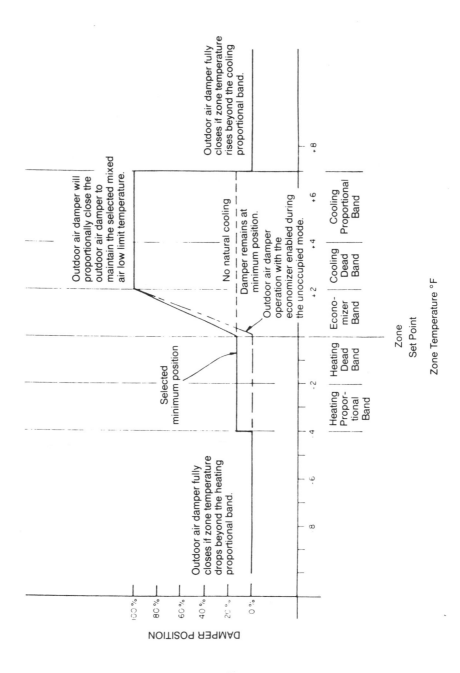

(6-13. Continued)

DDC AUTOMATION & DESIGN

A total automation system must possess a broad range of data gathering and automatic control capabilities. Such functions are essential for hvac/r, fire safety and building security monitoring systems. In addition, a carefully planned and executed energy management program must be implemented and systematically monitored. These complex functions can only be performed by a software-based control system.

■ AUTOMATION ARCHITECTURE

DDC plant and building automation systems are based on a concept of comprehensive control. All pertinent mechanical and electrical equipment is systematically monitored and operated to provide the optimum in design performance and operating efficiency. This involves more than just hvac/r equipment; it often involves control of equipment and systems unrelated to hvac/r such as:

- lighting and electrical equipment
- fire alarm, prevention, control and life safety equipment
- plant and building security systems

A DDC software-based automation system can sense, measure and respond within microseconds to any deviation from design conditions. This system can calculate variables and provide reset performance to compensate for changing load conditions or comfort requirements. It can also detect operating and energy-wasting abnormalities.

The software provides for either remedial adjustment or instant identification of trouble areas. This surveillance permits maintenance technicians to identify and correct potential problems before they develop into major difficulties. The DDC automation system provides the aforementioned capabilities by:

- continuously scanning all sensor points
- providing accurate indication of all sensed values
- logging all sensed values
- initiating alarms for problem areas
- fluctuating remote stop and start of equipment for energy maximization
- maintaining remote readjustment of control set points
- supplying verbal, visual and documented communications of all performance values

A basic automation system is shown in Figure 7-1. The core of the computer system is the CPU, which contains the central intelligence software required for system management and energy optimization. This configuration enables an operator to communicate with the entire automation network or networks. The CPU is also known as the host computer since it controls a multiple computer network and issues global commands for total system coordination.

In addition, the CPU has communication ports to interface with operator terminals, printers and screens. A modem provides remote communication by telephone. Trunks #1 and #2 in Figure 7-1 show a bus device which links the CPU with RPU's in a network configuration. These various components provide the man-machine interface necessary for accurate and effective operation.

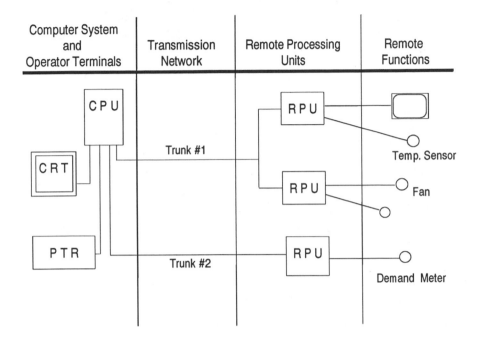

Computer System and Operator Terminals	Transmission Network	Remote Processing Units	Remote Functions

7-1. Basic Building Automation System (BAS)

RPU's are usually stand alone microcomputers that provide data acquisition for the CPU and sensors, and also provide output control for each operating zone. An RPU monitors input from sensors and makes decisions for equipment operation. It communicates exceptional conditions to the CPU and responds to the CPU's central direction. However, if the CPU fails, the RPU continues to independently manage conditions at its local level, as will all the RPU's in the network.

In effect, an RPU concentrates system intelligence for the zone being monitored. Within its sphere of operation, its software is programmed to provide all functions of automation, regulation, programming, recording and calculation required for precision control and performance modification for weather and other environmental variables. In addition, the RPU provides the locations for binary and analog sensor inputs, controlled device outputs, and set point and proportional band adjustments. However, despite the RPU's independence, all programming is first done on the CPU and then down-loaded to the stand alone RPU's.

■ BUILDING & PLANT AUTOMATION SYSTEM CONFIGURATION & DESIGN

The engineering of a DDC automation system must entail a well-planned, organized and systematic procedure. The computer is a powerful tool with unlimited capability if it is accurately programmed. However, it does not have the capability to supply missing details or correct program errors. Consequently, accurate computer functions depend upon system designers to attend to the most minute details and operating procedures during programming.

Figure 7-2 illustrates an organized five-step approach to system design and application. The first step involves studying the plans and specifications (specs) of the plant or structure. Usually the sales engineer or members of the DDC engineering staff have first access to plans and specs. Preliminary consultations with architects and design engineers can also provide pertinent details, which can be noted in worksheet #7720 (Figure 7-3). These can include owner's input and overall requirements as well as a listing of all hvac/r equipment and distribution systems to be controlled. If an **energy management system (EMS)** is required, equipment other than hvac/r (e.g., elevators, lighting, sprinklers, smoke detection, fire safety, entry control and intrusion detection) must also be described in detail.

The second step involves a question-and-answer session between members of the DDC engineering organization and any interested outsiders. Worksheet information and all other pertinent information must be analyzed to formulate a process code for the control system. The **process code** is a logical listing of all control routines and parameters required to provide the specified automation performance.

The following should be discussed during this phase:

- the number of building or plant systems and zones to be controlled

- what constitutes hvac/r equipment (e.g., chillers, refrigeration pumps, compressors, boilers, air handlers and cooling towers) and functions to be controlled

- the non-hvac/r control functions (e.g., lighting and electrical, fire detection and safety, and building security and surveillance)

- the energy management control requirements (e.g., load shedding, duty cycling, start/stop time, optimization and reset)

- the types of control sensors (e.g., digital/analog, pneumatic, electromechanical or solid state) and their signal characteristics

- types of control devices required (e.g., pneumatic, electromechanical or solid state) and their performance (digital or analog, 0-20, pneumatic, 4-20 mA current loop or 0-20V DC)

- the interface or transducer requirements

- the hardware requirements for DDC Automation systems (e.g., CPU's and RPU's)

- type and number of operator stations required (do stations need full interaction capability or only alarm and surveillance capability?)

- the future expansion projections for the system

When these and any other related questions have been answered and evaluated, plans can begin for complete system design.

I. FROM PLANS AND SPECS

TO DSC—DIGITAL SYSTEM CONTROLLER

QUICK SUBMITTAL

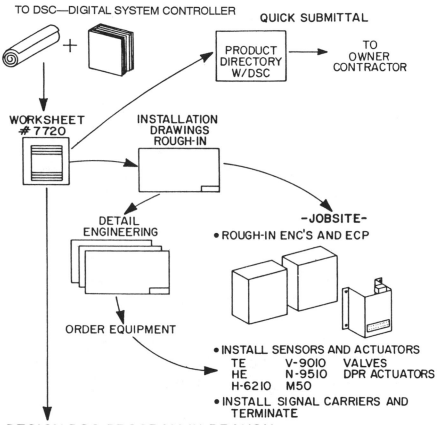

PRODUCT DIRECTORY W/DSC

TO OWNER CONTRACTOR

WORKSHEET #7720

INSTALLATION DRAWINGS ROUGH-IN

DETAIL ENGINEERING

-JOBSITE-

• ROUGH-IN ENC'S AND ECP

ORDER EQUIPMENT

• INSTALL SENSORS AND ACTUATORS

TE	V-9010	VALVES
HE	N-9510	DPR ACTUATORS
H-6210	M50	

• INSTALL SIGNAL CARRIERS AND TERMINATE

2. DESIGN DSC PROGRAM IN BRANCH

USE "DECISION TREE" Q'S AND A'S TO DETERMINE "PROCESS CODE"
• MANUALLY WITH APPLICATIONS MANUAL OR
• WITH LIBRARY OF TYPICAL SYSTEMS

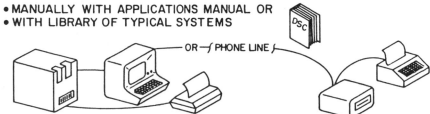

— OR —{ PHONE LINE }—

DSC

• FILL IN SOFTWARE DESIGN FORMS (FORMS 7721-7727)

7-2. Design Procedure for a Digital System
Courtesy of Johnson Controls, Inc.

3. GENERATE FLOPPY DISK CONTAINING CODE

USE WANG BRANCH TERMINAL AND FLOPPY LOADER DEVICE (FLD-IOI)
OR ⌐ PHONE LINE ⌐

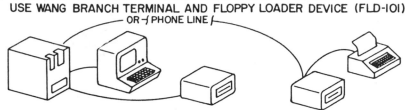

4. LOAD PROGRAM INTO FIC AND VERIFY ON BRANCH TEST PANEL

LOAD WITH FLD-IOI   BRANCH TEST PANEL

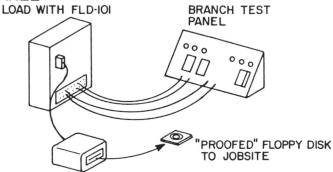

"PROOFED" FLOPPY DISK TO JOBSITE

5. CONFIGURE DSC ELECTRONICS IN THE ENCLOSURE AND COMMISSION DSC

CONFIGURE   COMMISSION

USE CDB FUNCTIONS TO TEST, RUN DIAGNOSTICS, SET GAINS AND SET POINTS

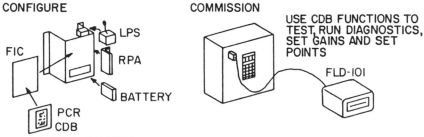

LPS
FIC
RPA
BATTERY
FLD-IOI
PCR
CDB

- PROGRAM CHANGES
  USE WANG BRANCH TERMINAL, IN-BRANCH OR OVER PHONE LINES, OR USE PSDU FOR LINE-BY-LINE CHANGES
- GENERATE AS-BUILT DRAWINGS AND "AS-BUILT" FLOPPY DISK
- CUSTOMER TRAINING AND JOB ACCEPTANCE (CERT. OF COMPLN. FORM # 4790)
- WARRANTY AND CONTRACTED PREVENTIVE MAINTENANCE

(7-2, Continued)

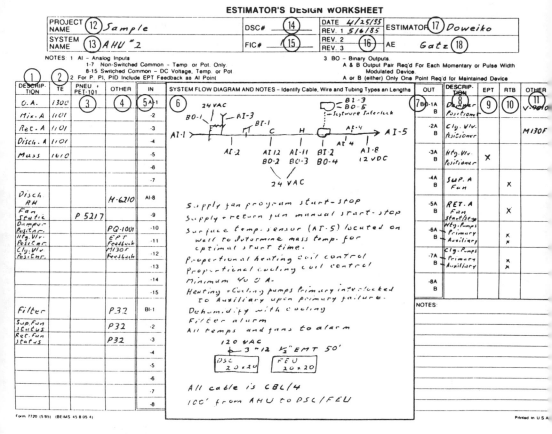

7-3. Form #7720 Estimator's Design Worksheet
Courtesy of Johnson Controls, Inc.

SYSTEM DESIGN ENGINEERING STRATEGY: POINT LIST

Designing a system involves creating complete lists of all points on the job. **Point lists** should include specific information on the type of point and field input. The points should be organized in tentative system grouping.

Point lists may be found in the specs and in the sales engineer's takeoff documents. If these are unavailable, create a point list by examining job correspondence and documents. However, this list should be double-checked for omissions, errors and late changes.

The point list (also called an I/O summary) contains all digital/analog and input/output points in the Building Automation System (BAS). Consequently, point selection is one of the most important functions of system design. System performance and return on investment are contingent upon the accurate selection and coordination of control points.

THE I/O SUMMARY

The following is a well-established control manufacturer's systematic procedure for developing input/output documents.

An I/O summary is a matrix of all digital and analog, input and output points incorporated in a BAS. Inputs may include input alarms, status, and analog values such as temperature, pressure, humidity and air movement. Outputs may include digital or analog commands for equipment operation or control point adjustments. I/O summary points may be either physically connected to the system (as with a sensor) or they may be software points resulting from a CPU calculation.

Selecting input/output points for listing in an I/O summary is typically more important than the accompanying prose description of the system. This is certainly true in cases where return on investment is an important consideration, as some points provide a high return relative to cost while others are marginal. The investment in hardware and wiring remote from the central console typically accounts for more than one half of total system costs.

Selecting I/O Points

To select I/O points, begin by asking:

- Which points should be physically connected?
- Which points should be calculated through software?
- Which points should be displayed or printed?
- Which points are consistent with system scope and return on investment?

To answer the above questions, it is helpful to divide points into three categories:

1. Points controlled (i.e., start/stop)

2. Points monitored (i.e., status)

3. Points for analysis, energy management and record keeping

1. POINTS CONTROLLED

Some points may be started or stopped (turned on or off) to save energy. Central control may also suggest ways to reduce manpower, reduce the need for additional manpower or make existing manpower more effective. This frees up more time for preventive maintenance.

Items commonly controlled, either manually or automatically:

Air handling unit fans

Exhaust fans - radiation pumps (on/off/auto)

Lighting - interior and exterior

Boilers

Chillers with associated pump

Electric reheat

Remote reset of local loop control setpoints and damper positioning

2. POINTS MONITORED

Usually alarms (digital or analog) are used to promptly alert personnel of hazards, failures or trouble conditions. Building personnel frequently become aware of a fault or danger before other building occupants. As a result they can often correct the problem before equipment damage or critical downtime occurs. In addition, selecting the proper alarm reduces the frequency of regular equipment room tours.

Items commonly equipped with an alarm:

Air Handling Units:

No air flow (fan status)

Fire alarm-fire stat or smoke detector

Low mixed air temperature

Filter (clogged and/or run out)

Discharge air temperature

Chiller Systems Alarms:

Chiller alarms, from chiller control panel

Condenser water pump status (flow)

Chilled water pump status (flow)

Condenser water temperature

Chilled water temperature

Low cooling tower sump temperature (freeze)

No water flow (pump)

Boiler Exchange Alarms:

Low steam pressure alarm

Temperature alarm, heater (hot water)

Domestic water temperature

Down stream pressure for PRV's

Boiler shutdown (taken from flame safeguard)

—Low water

—Ignition failure, etc.

Converter - supply and return water

No water flow (pump status)

Stack temperature

Fuel level

Miscellaneous Alarms:

Sewage ejector alarm

Sump alarm

Condensate receivers; using temperature and intercom

Snow melt

Expansion tank levels

Penthouse door open

Transformer vault temperature

Central air alarm (low/high)

Chemical feed

3. POINTS FOR ANALYSIS, ENERGY MANAGEMENT AND RECORD KEEPING

Points may be utilized to analyze the operation of a building's mechanical and electrical systems and to determine if they are operating as properly or efficiently as possible. Selected temperatures, pressures, relative humidities, KWh's and other analog values may be connected to provide building personnel with a means of evaluating the correct operation of a system.

Items commonly used for analysis, energy management and record keeping:

Return air temperature

Return air relative humidity

Zone temperatures (usually sample zones only; if the space stat is reset from the building automation system, then space temperature should be included)

Cooling coil discharge

Supply and return water from cooling coil

Preheat coil discharge

Outside air temperature

Outside air relative humidity or dewpoint

Pressure reducing valves, downstream pressures

Chilled water supply and return temperature

Condenser water—supply and return temperature

Supply and return water temperature from converters

Cooling tower sump temperature

Other items helpful in evaluating system operation:

Cooling tower fan speed operation indication (fast/slow/off)

Damper position indication (direct acting/reverse acting, positive feedback from damper)

Location of intercom pickups for audible system monitoring

Positive feedback of vortex damper positions

Fuel level indication

Items frequently calculated:

Degree days

Equipment run time

Enthalpy, outdoor and return air

GPM

Btu

kW and kWh

Flow

Efficiency

Differential temperature

Average temperature

Note: Historical trend logs are often printed daily, monthly and yearly.

Input/Output Summary

CHILLER SYSTEMS

System, Apparatus, or Area Point Description	Analog — Measured (Temperature, Pressure, RH, KW)	Analog — Calc. (KWH, Enthalpy, Run Time, Efficiency, GPM)	Binary (Status, Filter, Smoke, Freeze, Off-Slow-Fast, Hi-Lo)	Commandable Pos. (OFF-ON, OFF-AUTO-ON)	Commandable Grad. (Cntrl. Pt. Adj., Dmpr. Pos.)	Alarms (Hi Analog, Low Analog, Hi Binary, Low Binary, Proof, MAINTENANCE CRITICAL)	System Features / Programs (Time Scheduling, Demand Limiting, Duty Cycle, Start/Stop Opt., Enthalpy Opt., Reset, Event Program, DDC, Alarm Instruct, Maint. Work Order)	General (Intercom, Color Graphic)	Supplementary Notes
CHILLER SYSTEM #1		Run Time: X	Status: X					X X	
CHILLER	KWH: X	Run Time: X	Status: X			Proof / Maint. Crit.: X			NOTE 1.
COND. WTR. PUMP(S)		Run Time: X	Status: X	OFF-ON: X		Low Analog / Proof: X X			NOTE 2.
CHILL. WTR. PUMP(S)		Run Time: X	Status: X	OFF-ON: X		Low Analog / Proof: X X			NOTE 2.
COND. WTR. SUPPLY	Temperature: X					Hi/Low Analog: X X			
COND. WTR. RETURN	Temperature: X					Hi/Low Analog: X X			
CHILL. WTR. SUPPLY	Temperature: X	Efficiency: X			Cntrl. Pt. Adj.: X	Hi/Low Analog: X X			
CHILL. WTR. RETURN	Temperature: X					Hi/Low Analog: X X			
COND. WTR. HEADER	Temperature: X					Hi/Low Analog: X X			
CHILL. WTR. HEADER	Temperature: X X					Hi Analog: X			
COOLING TWR. FAN(S)	Temperature: X		Off-Slow-Fast / Hi-Lo: X X			Low Analog: X X			
COOLING TWR. SUMP						Maint. Crit.: X			
COOLING TWR. SUMP	Temperature: X								

Notes: 1. CHILLER SHUTDOWN ALARM TAKEN FROM CHILLER CONTROL PANEL
2. STATUS ON BOTH PRIMARY AND STANDBY REQUIRED

Page _____
Of _____
Form 1291 (7/82)
Printed in U.S.A.

7-4. Typical DDC Chiller System Input/Output Designations
Courtesy of MCC Powers

Input/Output Summary

System, Apparatus, or Area Point Description	Temperature	Pressure	RH	KW	KWH	Enthalpy	Run Time	Efficiency	Status	Filter	Smoke	Freeze	Off-Slow-Fast	Hi-Lo	ON-OFF	OFF-AUTO-ON	Cntrl. Pt. Adj.	Dmpr. Pos.	Hi Analog	Low Analog	Hi Binary	Low Binary	Proof	Time Scheduling	Demand Limiting	Duty Cycle	Start/Stop Opt.	Enthalpy Opt.	Reset	Event Program	DDC	Alarm Instruct	Maint. Work Order	Intercom	Color Graphic	Supplementary Notes
	Analog — Measured				**Calc.**				**Binary**						**Commandable — Pos.**		**Grad.**		**Alarms**					**System Features — Programs**										**General**		
OUTSIDE AIR	X		X			X																														MISCELLANEOUS
SNOW MELT									X																											
SUMP									X						X																					
SEWAGE EJECTOR									X												X	X														
EXHAUST FAN(S)									X X						X X						X															QUANTITY =
FOUNTAINS									X						X																					
LIGHTING CONTROL									X						X																					CONTROL CIRCUITS =
FIRE ALARM PANEL									X												X	X														ZONES =
EMERG. GEN.									X						X																					
FUEL OIL TANK									X												X															LOW LEVEL
FIRE PUMP																					X	X														
JOCKEY PUMP																						X														
DEMAND METER								X																												

Alarm section handwritten labels: MAINTENANCE / CRITICAL

Page _____
Of _____

Form 1291 (7/82)
Printed in U.S.A.

Notes:

7-5. Typical DDC System Input/Output Designations
Courtesy of MCC Powers

Typical I/O summary forms for a common air handling unit, a boiler system and a chiller system are shown in Figures 7-4 and 7-5.

ASSIGN POINT LOCATIONS

After listing and analyzing I/O point summary information, assign point locations in the software program. Figure 7-6 shows a form used to assign point locations for analog input points. Additional information needed for the computer should also be listed. Include data such as function identification, description of operating parameters and any other pertinent information. This form delineates the analog input sensor and all of its operating parameters in a format that can be programmed into the microcomputer.

For a complete automation program, fill out the following forms, listing functions and parameters similar to Figure 7-6: control point definition, analog data point definition, binary data point definition, control and display board function and control process definition.

GENERATE THE PROGRAM

Once the I/O summaries, point definition forms and other pertinent diagrams are completed, they are utilized by professional programmers to develop a specific operational software program. This information can even be fed into a computerized program generator which has been specifically developed to program a complete DDC system on a floppy disk for system application.

PROGRAM THE SYSTEM

After the DDC system program is constructed it is ready to be loaded into the host computer. The system manufacturer provides specific instructions for complete, step-by-step program loading. Also, it is very important to make backup copies of the program in case of future problems. Instructions for making duplicate backup copies are also provided by the

ANALOG INPUT POINT DEFINITION

PROJECT NAME		CONTRACT NUMBER	
SYSTEM NAME		DESIGNER	
DSC-8500 ADDRESS (1-32) ____	FIC ADDRESS (1-7) ____	DATE ____	PAGE ____ OF ____

PT. LOC	NAME	TYPE – NO.	P1	P2	P3	P4	P5	P6	P7	DESCRIPTION
AI-0	VCF	FUL – 1	D	0.0	E	N	O	0.0	1.0	VOLTAGE CORRECTION FACTOR
AI-1		–								
AI-2		–								
AI-3		–								
AI-4		–								
AI-5		–								
AI-6		–								
AI-7		–								
AI-8		–								
AI-9		–								
AI-10		–								
AI-11		–								
AI-12		–								
AI-13		–								
AI-14		–								
AI-15		–								

PT. LOC	NAME	TYPE – NO.	P1	P2	P3	P4	P5	P6	DESCRIPTION
AI-16		–							NOTE: AI-16 to -23 for "TOT" & "RAT" Points Only
AI-17		–							
AI-18		–							
AI-19		–							
AI-20		–							
AI-21		–							
AI-22		–							
AI-23		–							

7-6. DDC System Analog Input Identification
Courtesy of Johnson Controls, Inc.

manufacturer, and essentially entail using the DOS command DISKCOPY.

To load the program, first gain entry to the microcomputer (through the programmer's panel or terminal) by using a password. Although it is possible to log on the program by powering the system, it is not possible to gain access to the system without entering an established password. This procedure insures that only management-authorized individuals have access to the system.

The next step is to interface with the plant and building automation system through the available I/O devices such as the keyboard, CRT display terminal and printer. These devices provide the ability to communicate with all other microcomputers in the network and to note the current conditions of every point in the system. These points are maintained in the CPU memory by an operating system sub-routine and are periodically polled.

The rate at which a specific point is polled depends upon the number of devices in the system, their configuration, their relative importance and the rate of transmission (baud rate). Generally a significant and important change-of-state (COS) in any field device is detected by the CPU within seconds of its occurrance.

The DDC automation system is a collection of software programs and routines which provide housekeeping, communication, scheduling and timing involved in the effective application and operation of field hardware. These routines are supported by an executive program which involves input/output, scheduling, timekeeping and power failure backup services.

The Operating System Program Control communicates with field hardware, operator devices, process system outputs and command requests. It initializes software under restart conditions following power failure and directs the memory usage of applications programs. The many software subroutines which make up the DDC Automation Operating System fall into seven major categories:

1. Operator devices: provide the terminals and displays which allow interfacing with the automation system.

2. Field communications: handle communication between the CPU and all field processing units (FPU) in the system network.

3. Change of State (COS) processing: handles data responses from network units which indicate Changes-of-State occurring in field devices. This process can involve alarm operations requiring operator acknowledgement.

4. Commands: can be initiated by the operator or by the program. These are scheduled and executed by the Command Processor sub-routine.

5. Time Manager: is a timekeeping function which allows the operating system to set up required timing functions.

6. Systems Initialization: restores the operating system after power failure or prolonged outage.

7. Systems Modification: enables the operator to modify or add data base parameters.

CONFIGURE DIGITAL SYSTEM CONTROLLER (DSC) AND COMMISSION DDC SYSTEM

After finalizing design and programming steps, plan the physical placement and housing connection of the hardware components in the automation system (see step #5 in Figure 7-2). This process is known as **system configuration**. Once system planning and installation is complete, place the DDC in operation and check it for errors and malfunctions. This is called the **commissioning** stage.

System Configuration

The system configuration phase involves six major steps:

• Primary Processing
• Distributed Processing

- Field Processing
- Sensing Control
- Communication/Transmission
- Operator I/O

Because system configuration is of primary importance, each manufacturer outlines a very thorough and specific set of instructions in order to establish and maintain an ideal working environment for all system hardware.

System Maintenance And Diagnosis

Automation system manufacturers usually supply a comprehensive software package for an organized and systematic maintenance program. A package of this nature can automatically generate and schedule work orders, provide historical information on past maintenance activities and prepare all-inclusive reports for management evaluation. These functions provide plant and building managers with an accurate system perspective on work schedules and values of labor, material and inventory.

Implementing a maintenance program involves six basic phases:

1. Data base entry

Pertinent information to be entered in the computer includes:

a. Equipment information (e.g., equipment name,description, manufacturer, model number, location)

b. Preventive maintenance routines (e.g., job number, estimated time, inventory, worker's skill level, priority, reschedule interval, instructions)

c. Inventory (e.g., part number, stock number, description)

d. Skill level information

2. Automatic scheduling and generation of work orders

3. Work actually performed

4. Input of completion form

Upon completion of the work order, the pertinent information is entered in the computer by answering specific questions provided by the program menu.

5. Generation of reports

6. Analysis of reports

Although software maintenance programs minimize and prevent problems, malfunctions can still occur. As a result, **microprocessor-based test units** are used to assist software programs. These supplementary devices help localize and identify problems in the system.

The test unit can be interfaced with the system control panel. When connected, it emulates job-site hardware for comparison with communication responses received at the master device. In addition, the unit can monitor communication activity and also function as a master control to verify operations and connections.

■ SUMMARY

Modern technology has provided the means for complete plant and building automation. Automation makes the "management by exception" goal of upper-level academicians attainable. Consequently, human intervention is typically only required when occasional problems arise that cannot be remedied by the system.

AFTERWORD

Over the years, control systems have improved considerably. They have evolved from manual operation to fully automated, highly complex computer-based systems. However, knowledge of fundamental control theory is always important whether the application is residential or commercial, simple or complex. For example, simpler systems are more suitable and cost-effective in many applications. And in the more sophisticated applications, a solid knowledge of the basics is necessary for proper system design, engineering, application and maintenance.

Hvac/r system automation and surveillance was made possible by the advent of the microcomputer. Devices such as the programmable controller and analog/digital conversion units enable the computer to process sensor signals and carry out specialized hvac/r functions with little supervision.

Previous knowledge of computer language, hardware or software is not required to design, implement or operate a microprocessor–based control system. System programming is simplified by matching ladder–logic diagrams to the manufacturer's symbols included with the PC. In this manner, even highly complex computer-based systems are accessible ot all hvac/r professionals. Remember, all technology, no matter how self–sufficient, depends upon Man for accurate design, implementation and operation.

GLOSSARY

For communication to be effective, the symbols, abbreviations and terms must be accurately interpreted. This is doubly true in high-technology areas, especially those involving computers. Since the introduction of DDC, control industry communications—both verbal and written—have involved an unusual amount of acronyms, abbreviations and industry–exclusive expressions. Consequently, this glossary is provided as a reference for terms used in the control industry and in this text.

Accessory Unit

An optional add-on module offering additional functions (e.g., indicators for temperature, relative humidity, alarms, pilot lights) not provided by the input or output modules of a computer.

Access Time

The amount of time it takes a computer to obtain data from an address.

Address

Numeric code which identifies the devices in a digital system to the controller; specific location for every bit of data entered Address Selector Switches on the controlled device which are used to assign data an address code number.

Algorithm

Prescribed rule or process to be followed for the solution of a specific problem. Algorithms solve problems mathematically using a finite number of steps.

Analog

Data with continuously variable physical quantities (e.g., pressure, temperature, humidity, air flow, voltage, resistance). These variables can be processed by the computer in accordance with the mathematical formulas programmed in it.

AND Gate

A gate circuit (in a computer) which has more than one control input terminal; no output signal will result unless a pulse is applied to all inputs simultaneously.

Analog Input & Output

An input or output of variable amplitude or value which reflects variations of sensory devices (e.g., temperature, pressure, humidity, voltage, resistance).

Analog-to-Digital Converter (ADC)

A device which converts sensed input information in proportional or modulating (analog) form to digital form, which provides a measurable timed pulse whose duration is directly proportional to the analog input.

Architecture

The configuration of hardware in a DDC system.

Array

An organized set of elements in a computer program. Arrays can be referenced singly or collectively by using the array name and one or more subscripts.

Asynchronous Communication

Data transmission in which the time interval between characters varies.

Automation

Total automatic operation of environmental energy management, fire alarm and safety, and security systems in a building.

BASIC

Beginner's All-Purpose Instruction Code—a commonly used computer programming language.

Binary

Two-position data (either on or off) based on the binary number system. It uses 2 as its base, which only uses the digits zero (0) and one (1).

Baud Rate

A unit of signaling speed equal to the number of signal events or code elements per second and based on the duration of the shortest element. For example, if each element carries one bit, the baud rate is numerically equal to bits per second (bps).

Bit

Acronym for binary digit, the smallest unit of memory in a computer. It is in one of two states represented by the binary digits 0 and 1.

Break

A method or signal to stop program execution. Some computer keyboards have a key labeled "BREAK" which stops a program in progress.

Boolean Algebra

A mathematical logic system named after George Boole. Boolean algebra uses binary digits (0 and 1) to delineate logic decisions. The logic symbols are used to design and program microcomputers.

Buffer

An intermediate storage area between two devices which compensates for a difference in access time or format during transmission of information.

Bug

A program defect or error.

Building System

An energy system or building automation system (BAS) which governs or protects the overall building operation. It can include systems regarding environmental factors, building security, protection and power failure.

BUS

A set of parallel conductors or wires used as paths for information transmission.

BUS, Bi-directional

Conductor or wire used for two-way digital communication between the controller and controlled devices.

Byte

The smallest addressable unit of memory that the computer processes as a group. A byte is a group of eight bits. It can represent one character or two numerals. Memory size can be related to the number of bytes.

Cable, Twisted Pair

Two thin conductors are formed into a cable by twisting, which reduces intercapacitance.

Calibration

Adjustments which make an instrument reading or desired output match actual output.

CAL Language

Custom Application Language. Programming language used by design engineers and usually involving four elements: field points, data points, panel functions and control processes.

Card

Printed circuit wiring and component unit which can be plugged in internally.

Cathode Ray Tube (CRT)

A tube containing an electron beam which produces a visible image (e.g., a television picture tube).

Centralized Control

A system of data analysis and control action performed at a central location.

Central Processing Unit (CPU)

This computer unit interprets and executes program instructions. It does not include input/output functions or memory.

Characterization

Information given to the RPU that completely describes its operational characteristics (e.g., how many physical points are on each function card, the range of each analog-input, or which function cards are in which slots).

Checkpointing

A procedure for swapping data between main memory and disk memory.

Chip

A single device consisting of transistors and diodes interconnected chemically and usually cut from a silicon wafer. A chip can contain all elements of a central processor.

Clock

The timing portion of the microprocessor.

Compensating Sensor

The system detector which senses a variable other than the controlled variable and resets the main sensor's control point.

COBOL

COmmon Business Oriented Language, a computer language that makes use of English language communications.

Compensation Ratio

The number of unit changes in the compensating variable which produces one unit change in the controlled variable control point. Divide the authority of the compensating variable into 100 to get compensation ratio.

Console

The computer component where the control keys and input/output devices are located (e.g., printers, CRT's).

Controlled Device

Device actuated by a controller to perform the control function.

Controller

A control device which maintains the predetermined output of the equipment it is controlling.

Control Hunting

Control system instability in which the controller over-corrects in each direction from the set point.

Control Point

1. The actual point at which the system is controlling pressure, temperature, humidity pressure or air flow; may not always be the same as set point.

2. In a microprocessor, a software point which links hardware inputs to hardware outputs.

Control Unit

Sometimes referred to as the control section of a digital computer, it is the section which directs sequence of operation and sends proper signals to other computer circuits to carry out instructions.

Core Memory

The memory device in a computer which contains magnetic cores, tiny units of magnetizable metal that can be either in an on or off state (thereby representing either a binary 1 or 0, respectively).

Crosstalk

Undesired energy in one signal path (disturbed circuit) as a result of coupling from another signal path (distributing circuit). Also called Cross Coupling.

Counter

A circuit designed to count input pulses.

Cyclic Redundancy Check

A means of detecting transmission errors generally between a computer and a remote processing unit (RPU).

Daisy Chain

A network of microprocessor units interconnected in such a manner that signals pass from one to another in serial fashion. Each unit can modify the signal before passing it on to the next device.

Data

A term which denotes all I/O facts, information, numbers, letters and symbols processed by a computer (e.g., integer numbers, single precision numbers, double precision numbers, character sequences or strings).

Dead Band

A range in which a control signal may vary without initiating control action.

Debug

The process of locating and removing logical or syntactic errors from a program. Also a method of detecting and correcting malfunctions in the computer itself.

Decibel

A standard unit denoting a loss or gain in transmission power.

Dedicated

A term for equipment, programs or procedures which are designed for specific use or application (e.g., microprocessors designed for games, calculators, control applications).

Default Value

A value for a variable which the controller selects if a specific value is not selected.

Degauss

A method of demagnetizing magnetic tape.

Delimiter

A computer character which limits a string. It marks the beginning or end of a data item but is not a part of the data or string.

Derivative

Rate at which something changes, or quantities increase or decrease. Also the name of a control action in which the output is proportional to the input's rate of change.

Digital

Discrete integral numbers to express information stored, transferred, transmitted or processed by a dual procedure (true/false, on/off, open/closed).

Digital Data Point

An "either/or" (true/false, on/off, open/closed) remote point which can transmit data to the controller.

Digital Input/Output

Input or output information converted into binary numbers for processing by the digital computer.

Direct Action

An increase in the controlled variable creates an increase in the controller's output.

Direct Digital Control

A system in which digital computer outputs are used to control a process, and the monitored signals are fed back to the computer in digital form. The computer can compare variable analog inputs, perform the control mode calculations and output a digital signal to the control devices.

Disable

The interruption or prevention of a function.

Distributed Control System

A network of independent, stand alone control units under the jurisdiction of a central computer but having individual capability to analyze data and initiate control action.

Down Line Loading or Down Loading

Transmitting information to a remote processing unit (RPU) from a computer for program loading or characterization.

Dump

Accidental or intentional withdrawal of power from a computer; also the transfer of computer memory from one section to another.

Economizer Control

In air handling systems, this is the control system which selects either outside air, recirculated air or a mixture of both for the most energy-efficient cooling.

Enable

A state which allows a function to occur.

Energy Management System

An organized system which maximizes total building structure and equipment for optimum energy efficiency and cost effectiveness.

Enthalpy Control

A method of sensing total heat to determine the maximum cooling capability of two possible air sources.

Entry Point

The address of a machine language program or routine where execution is to begin.

EPROM

Erasable Programmable Read Only Memory. Memory capable of being electrically programmed apart from the computer. It functions as permanent Read Only Memory, yet it can be erased by exposure to ultraviolet light.

Error Signal

In control technology, a signal by the controller which represents a deviation from set point. It is measured and used to correct the deviation by re-positioning a controlled device.

Exit

A method used to halt a repeated computer cycle of program operation.

Feedback

An action in which the ouput response of the controlled device is fed back to the controller.

Floppy Disk System

Floppy disks provide random access to program and data storage, and the host computer can readily record data on or retrieve data from the disk.

Flow Chart or Flow Diagram

Contains all the steps in a computer program; graphically depicts the sequence of operation using symbols, directional marks and accepted representations to indicate the steps in computer operation.

FORTRAN

The name is a contraction of FORmula TRANslation. It is a programming language which allows the programmer to state in simple terms the procedural steps to be carried out by the computer.

Gain

The ratio between an output signal and the input signal.

Gate

A circuit whose output is deenergized until specific input conditions have been met. Properly pulsing a gate circuit can trigger passage of other signals through a circuit.

Graphic File

A storage location for operator instructions on how to create pictures on a color CRT.

Handshake

Descriptive buzzword for the process of buffering or interfacing peripheral computer units or programs.

Hardware

Refers to the physical or tangible computer equipment.

Hard-wired

Circuits wired for specific purposes and, in effect, unalterable.

Heat Sink

A device typically involving thermal mass which is used to dissipate heat from heat-vulnerable components.

High Level Language

A problem- or procedure-oriented language (as opposed to machine- or memory-oriented language).

Homeostasis

A system condition where input and output are precisely balanced; a steady state without change.

Host Computer

The primary or controlling computer in a network computer operation.

Hunting

Continuous overshooting and undershooting of a control system with the inability to achieve equilibrium (usually the result of controller over-sensitivity).

Hysteresis

The difference between the response of a unit and the increasing or decreasing force or signal applied.

IC Memory System

A system requiring very little support electronics because all sensing, decoding and driving circuits are built in.

Increment

The value added to a counter for each cycle of a completed repetitive procedure.

Incremental Control

Based on the ouput signal change for both velocity and acceleration to change in the controlled variable (similar to a floating control routine). As a result, it has a quicker response to load changes and provides a narrower dead band.

Initialize

A computer start-up procedure for loading and defining an operation.

In-circuit Test System

A circuit connected to all devices which provides fault diagnosis.

Input Register

A computer's internal storage register which accepts information outside of the computer at input speed and supplies it to the calculating unit with greater speed.

Input

A signal or data sent to or impressed upon a device or logic element.

Input Module

An input device or devices used to send data to another device. Most input modules have a DC voltage output which is proportional to deviation from set point.

Integral Control

The control output is proportional to the time integral of the error signal; the rate of change of the output is proportional to the input. This method uses the sum of system error over time.

Integrated Circuit

An interconnected array of logic components fabricated in a single semiconductor crystal.

Interface

A means of linking adjacent or unlike components (e.g., equipment, circuits, I/O devices) to provide the communication needed for complete control functions.

Jack

A socket or connecting device with attached circuit wiring which can be accessed by inserting a plug.

Keyboard

There are three basic keyboard types: *alphanumeric*, used for word-, text-, data- and tele-processing; *numeric* only, used on touch tone telephones, accounting machines and calculators; and *mixed* keyboard, a combination of the alphabetic and numeric keys.

Latching

An arrangement which holds a circuit in position.

Library

A collection of standard programs and sub-routines which can solve problems or parts of problems.

Limit Sensor

The sensor which overrides the main sensor when a predetermined limit has been reached.

Linear

A response in which output varies directly with input (e.g., each degree of temperature change results in a pressure change of one pound). This is depicted as a straight line when graphed.

Line Printer

A high-speed printer capable of printing an entire line at a time.

Load

The storage of information in a computer from auxiliary or external sources.

Local Loop Control

The controls for a system or sub-system that continue to function when the host computer or central computer is inoperative.

Log

A collection of information and messages that documents pertinent facts of equipment operation.

Logic

In microprocessors, logic uses symbols to represent quantities and relationships identifiable as gates or switching circuits; these can be arranged to perform logical functions necessary for equipment operation.

Machine Language

Refers to binary language and the ultimate language all computers must use. All other program languages must be translated into machine language before being executed by the computer.

Magnetic Core

A magnetic material which, when placed into one or two magnetic states, will remain in that state until changed by external action. It can provide memory storage, gating or switching in a computer's CPU.

Main Frame

The main parts of the computer which are basic to its operation.

Matrix

An array of quantities in a predetermined form which can be used as a coding or decoding network in a computer; a network of input and output leads with logic elements connected at intersections.

Memory

One of the basic components of a central processing unit which stores information for future retrieval.

Microcomputer

A smaller computing system similar to a minicomputer but with less speed and capability. Usually consists of a microprocessing unit on a printed circuit board and has memory and auxiliary circuits, and is capable of executing software.

Microprocessor

The central processsing unit (CPU) of a microcomputer; this section generally contains the arithmetic logic unit and the control logic unit.

Microsecond

One millionth of a second.

Minicomputer

A mainframe, usually a parallel binary system with either semi-conductor or magnetic core memory from 4K to 64K words of storage and a cycle time of 0.2 to 8 microseconds or less. It can consist of only one PC card but even then has higher performance than a microcomputer.

Mnemonic

A technique to assist a programmer's memory—mnemonics is a method of improving the memory efficiency of computer storage.

Modem

Derived from MODulation/DEModulation—a chip or device to change digital information to and from analog form. This enables computers and terminals to communicate information over telephone circuits. On the transmission end, the modulator converts signals or pulses to the right code for alternating current transmission over a telephone line. In receiving, a demodulator reconverts the signal to direct current to the computer's interface device.

Module

A packaged operational hardware unit, device or program unit which is discrete, self-contained and designed for use with other components.

Mole

A substance whose molecular weight is in grams—gram molecule (symbol: mol).

MOS Memory

Metal Oxide Semiconductor computer memory (instead of magnetic core memory).

Multidrop

Digital communication whereby all digital devices are connected in parallel to a common bus.

Multiple Loop Control

A single controller operates a number of independent systems.

Multiplex

The use of one channel for simultaneous transmission of more than one signal on a single channel.

Nano

Prefix for 10-9, a nano-second is one-billionth of a second.

Nanoprocessor

One that operates in the nanosecond cycle range.

Network

A combination of elements, devices or microprocessors which are interconnected with each other and with a host computer.

Noise

Unwanted electrical or transmission disturbance (voltage, current or data) In an electrical or mechanical system.

Nonlinear

Output and input which do not vary directly and cannot be graphed as a straight line.

Normally Closed

A controlled device which returns to the closed position when the control signal or power is removed.

Normally Open

A controlled device which returns to the open position when power or a control signal is removed.

Null

Zero output, the result of a balanced condition.

Numeric Code

A system of abbreviation whereby information is prepared for machine acceptance. Information is converted into numerical (not alphabetical) quantities.

Open Loop

A control system whereby outputs are controlled exclusively by system inputs; the system has no feedback from output to input, and as a result there is no self-correcting action (as in a closed loop circuit).

Output

The transfer of data from internal storage of a computer to external output devices (e.g., disk file, line printer, or external storage).

Overshoot

Where a controlled variable or device extends beyond set point equilibrium position. Hunting results when a control system overshoots and undershoots constantly (usually the result of oversensitivity).

Parallel Bus

A means to transmit simultaneous information bytes with the use of multiple conductors and ground connections.

Parameter

A set of specific values which are assigned to a control routine and which dictate how the control routine will operate.

Phase Angle

A measure of time whereby an output leads or lags an input. In electrical circuitry it is measured in degrees for sinosoidal variation.

Picosecond

One thousandth of a nanosecond.

Printed Circuit

A circuit with no conventional interconnected wires; instead, conducting strips are etched or printed onto an insulating board.

Process Control

The automation of continuous operations.

Program

A set of instructions (generally called software) which control procedures or processes, or solve problems.

Prompt

A message or character in a program which indicates readiness to accept keyboard input.

Programmable Control

A control system (generally for industrial equipment) which can perform the functions of conventional controls, (e.g., relays, timers, counters, drum timers, and step switches) and can be readily programmed directly from a ladder diagram.

Proportional

Control action which maintains constant linear relationship between the final control device and the signal from the controlled variable.

Proportional Band (Throttling Range)

The amount of change in the controlled variable which is required to drive the controlled device or actuator from one extreme to the other.

Protocol

A formal set of rules or conventions governing the format, timing and content of messages transmitted between two communicating processes or devices.

Pulse

The variations or change, rise and fall ,or finite duration in a quantity (e.g., electrical potential) having normally constant value.

Ramp

Sloping depiction of control performance.

Random Access Memory (RAM)

Provides access to any storage point in memory by means of vertical and horizontal coordinates; information can be stored or recovered in any order at random.

Read Only Memory

Memory of stored data which can be read but not changed.

Redundancy

The duplication of devices or features, each accomplishing the same function in order to asssure reliability of operation (e.g., wearing both suspenders and a belt).

Register

A digital computer device capable of containing one or more bits or words.

Relay Ladder Programming

A method of programming the programmable controller directly from existing relay ladder diagrams.

Reset

To restore a device to its original state. In computer jargon, to restore a binary cell to its initial zero position; also to return a storage location or register to zero or its initial condition.

Response Time

Time difference between generation of an inquiry and the receipt of the response; also involves transmission time to and processsing time from the computer.

Reverse Action

An increase in the controlled variable which causes a decrease in controller output (and vice versa).

Robot

A machine or device that operates itself independently, it can receive input signals or environmental conditions with sensors and perform prescribed and appropriate actions from stored programs.

Robotics

Areas of artificial intelligence performed by robots.

Routine

A set of computer instructions assembled in correct sequence for the purpose of directing the computer to perform the desired operation or operations. Each routine consists of a unique set of parameters or process variables.

Run

The linking of one or more routines to form one operating program.

Sensitivity

The minimum input signal which initiates a change in the controlled variable (e.g., as proportional band is narrowed, sensitivity is increased).

Serial Bus Communication

Each bit of information is transmitted consecutively through a single transmission line.

Servomechanism

A closed loop system whereby a small input power source controls a larger output power source.

Setpoint

The setting of the main sensor in a control loop which maintains the desired value of the quantity being controlled.

SI (Metric) Units

System International Units. Scientific international system of seven basics: length—meters; mass—kg; time r—in seconds; electric current I—ampere; temperature T—kelvin; amount of substance M—mol; Luminous intensity I—candela cd.

Slave

Electronic, pneumatic, or electro-mechanical unit or device controlled by signals from master equipment.

Smart Terminals

Self-contained terminals with human—oriented inputs and outputs.

Software

Non—tangible computer involvements—computer programs, languages, and procedures (as differentiated from physical hardware components of the computer itself).

Source Code

A computer program as written by the designer or programmer.

Stand Alone System

A control system microprocessor which can perform all basic control functions independent of the host computer.

String

A sequence of characters which are positioned according to a rule and must be examined verbatim for meaning.

Subroutine

The portion of a routine which can perform a specific task such as a mathematical or logical operation.

Subsystem

A self-contained portion of a system which can perform some major part of the system's functions.

Supervisory Control

A system which can exert supervisory corrective action to a stand alone system or independently operating control loops.

Syntax

Rules (grammatical) or relationships required in a statement or command for proper interpretation by a computer.

System

A collection of devices or parts which are organized into a whole by their interaction and interdependency with each other.

Transducer

A device which can convert energy flow from one transmission system to another (e.g., pneumatic to electric, electric to mechanical, etc.).

Twisted Pair

A digital circuit communication line whereby two insulated wires are twisted together but do not have a common covering.

Universal Asynchronous Receiver Transmitter

A specialized circuit module to convert serial transmission to parallel for microcomputer entry.

ABBREVIATIONS

ACC	Accessory
AD	Analog Data
ADC	Analog-to-Digital Converter
AI	Analog Input
ALU	Arithmetic Logic Unit
ASCII	American Standard Code Information Interchange
ATC	Automatic Temperature Control
BAS	Building Automation System
BBC	Battery Backup Clock
BD	Binary Data
BO	Binary Output
BPL	Boiler Profile Log
BPS	Bits Per Second
CAD	Computer Assisted Design
CAI	Computer Assisted Instruction
CAL	Custom Application Language
CAM	Content Addressable Memory
CDB	Control and Display Board
CEK	Control Electronics Kit
CIB	Communications Interface Board
COV	Change of Value
CP	Control Port
CPA	Control Point Adjustment
CPL	Chiller Profile Log
CPS	Characters Per Second
CPU	Central Processing Unit
CRT	Cathode Ray Tube
DAC	Digital-to-Analog Converter
DCCP	Duty Cycling Control Processing
DCK	Distributed Communication Kit
DDC	Direct Digital Control

DI	Digital Input
DO	Digital Output
DPU	Distributed Processing Unit
DSC	Digital System Controller
EAO	Electronic Analog Output
ENC	Standard Enclosure
EO	Enthalpy Optimization
EP	Equation Processor
EPT	Electric-to-Pressure Transducer
EPROM	Erasable Programmable Read Only Memory
FACE	Field Alterable Control Element
FAI	Frequency Analog In
FAP	Field Annunciation Panel
FEU	Field Equipment Unit
FEP	Front End Processor
FIC	Field Interface Control
FIFO	First In First Out (Memory)
FLD	Field Loading Device
FPU	Field Processing Unit
GIGO	Garbage In Garbage Out
HMA	Hardware Mounting Assembly
ICM	Intercom Station (remote)
I/O	Input/Output
ISAM	Index Sequential Access Method
LDI	Logical Digital Input
LDO	Logical Digital Output
LIFO	Last In First Out
LPS	Line Power Sub-assembly
MCR	Monitor Console Routine
MMI	Man/Machine Interface
MOS	Metal Oxide Semiconductor
MOV	Metal Oxide Varistor
MPL	Master Point List
MPLX	Multiplexer
MTO	Material Transfer Order
NLC	No Load Current
OAS	Operator Access Security

OG	Outboard Gear
OTT	Orion Temperature Transmitter
P	Proportional Control
PA	Pulse Accumulator
PAI	Process Analog Input
PAT	Powers Audio Trunk
PC	Printed Circuit Programmable Controller
PCB	Printed Circuit Board
PCR	Processor Board
PDL	Peak Demand Limiting
PET	Pressure-to-Electric Transducer
PI	Proportional + Integral
PID	Proportional + Integral + Derivative
PO	Pneumatic Output
PWM	Pulse Width Modulated
RAM	Random Access Memory
RCU	Remote Control Unit
RFC	Remote Function Circuit
RFG	Remote Function Guide
RLM	Relay Latching Module
RMM	Relay Momentary Module
ROM	Read Only Memory
RPA	Regulated Power Assembly
RPU	Remote Processing Unit
RTB	Relay Termination Board
RTD	Resistance Temperature Device
SCOV	Significant Change of Value
S/W	Software
SSTO	Start/Stop Time Optimization
TE	Temperature Element
TTA	Trunk Terminal Adaptor
TTL	Transistor-Transistor Logic
UART	Universal Asynchronous Receiver Transmitter
UPS	Uninterruptable Power Supply
VDU	Video Display Unit

INDEX